Collins

KS3
Science
Year 9
Workbook

Ian Honeysett, Sam Holyman and
Lynn Pharaoh

About this Workbook

There are three Collins workbooks for KS3 Science:
Year 7 Science ISBN 9780008553722
Year 8 Science ISBN 9780008553739
Year 9 Science ISBN 9780008553746

Together they provide topic-based practice for all the skills and content on the Programme of Study for Key Stage 3 Science.

Questions for each topic have been organised into sections that test different **skills**.

- Vocabulary Builder
- Maths Skills
- Testing Understanding
- Working Scientifically
- Science in Use

Found throughout the book, the **QR codes** can be scanned on your smartphone. Each QR code links to a video working through the solution to one of the questions or question parts on the double-page spread.

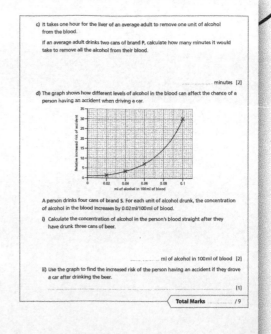

Track your **progress** by recording your marks in the box at the end of each skills section and the summary box at the end of each topic.

The **answers** are included at the back of the book so that you can mark your own work.

If you get a question wrong, make sure you read the answer carefully so that you understand where you went wrong.

Helpful tips are also included.

Contents

Biology

Chemistry

Physics

Answers

Acknowledgements

The authors and publisher are grateful to the copyright holders for permission to use quoted materials and images.

All Images are ©Shutterstock.com or ©HarperCollinsPublishers

Every effort has been made to trace copyright holders and obtain their permission for the use of copyright material. The authors and publisher will gladly receive information enabling them to rectify any error or omission in subsequent editions. All facts are correct at time of going to press.

Published by Collins
An imprint of HarperCollinsPublishers
1 London Bridge Street
London SE1 9GF

HarperCollinsPublishers
Macken House
39/40 Mayor Street Upper
Dublin 1
Ireland
D01 C9W8

© HarperCollinsPublishers Limited 2023

ISBN 9780008553746

10 9 8 7 6 5 4

Publisher: Clare Souza
Commissioning: Richard Toms
Authors: Ian Honeysett (Biology), Sam Holyman (Chemistry) and Lynn Pharaoh (Physics)
Project leader: Katie Galloway
Copyediting: Charlotte Christensen
Cover Design: Kevin Robbins and Sarah Duxbury
Inside Concept Design: Sarah Duxbury and Paul Oates
Text Design and Layout: Contentra Technologies
Production: Emma Wood
Printed in the United Kingdom by Ashford Colour Ltd

MIX
Paper | Supporting responsible forestry
FSC™ C007454

This book contains FSC™ certified paper and other controlled sources to ensure responsible forest management.

For more information visit: www.harpercollins.co.uk/green

Biology

Variation for Survival

Vocabulary Builder

1 Draw lines to join each genetic term with its correct description.

Genetic term	Description
DNA	One of 46 lengths of genetic material in a human cell
Chromosome	A length of genetic material coding for one protein
Gene	The chemical groups that hold the strands of DNA together
Base	The chemical that makes up the genetic material

[3]

2 This question is about heredity and evolution.

For each term, put a tick to show if each statement is **true** or **false**.

a) **DNA**   True   False

   Is found in the nucleus   ☐   ☐

   Is made of protein   ☐   ☐

   Contains two types of bases   ☐   ☐   [3]

b) **Species**

   Contains individuals that all look identical   ☐   ☐

   Contains organisms that can interbreed if they are
   of the opposite sex   ☐   ☐   [2]

c) **Continuous variation**

   Describes the range of heights of humans   ☐   ☐

   Is not affected by the environment   ☐   ☐   [2]

d) **Natural selection**

   Explains how evolution may have occurred   ☐   ☐

   Is happening now   ☐   ☐   [2]

3 Use words from the box to complete the sentences about variation and natural selection.

single	two	23	46	characteristics
cytoplasm	die	environment	evolution	
mutations	nucleus	survive		

All individuals show variation. The type of variation caused by the _____ is not passed on to the next generation.

Variation can also be caused by random mistakes when DNA is copied. These mistakes are

called _____. This type of variation can be passed on because genetic information is

carried in the _____ of every sex cell.

The sex cells contain _____ copies of every gene. This means that a fertilised egg cell

receives _____ copies of every gene.

Variation is needed for _____ to occur by natural selection. The best suited

individuals are more likely to _____ and reproduce. This means that they will pass

on their _____ to the next generation. [8]

Total Marks _____ / 20

Maths Skills

1 Two students want to investigate the variation in the width of ivy leaves in two different areas. One area is in the shade. The other area is in the open.

The students measure the width of 40 leaves in each area. They put their results into width groups. The results are shown in the table below.

| Width of leaves (mm) | Number of leaves in the group | |
	In the shade	In the open
0–19	4	6
20–39	20	8
40–59	7	11
60–79	5	8
80–99	4	

a) Calculate how many leaves were between 80 and 99 mm in the open area.

Write your result in the table. [1]

b) The students plotted the number of leaves in the shade in each width group on a bar chart.

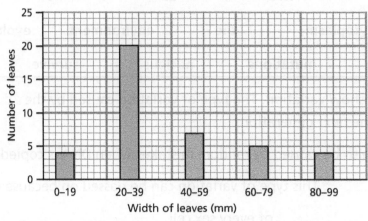

Legend: ▨ in the shade ▢ in the open

Y-axis: Number of leaves (0 to 25)
X-axis: Width of leaves (mm) — 0–19, 20–39, 40–59, 60–79, 80–99

i) Complete the graph by drawing bars showing the number of leaves in the open area in each width group. [2]

ii) Which width group in the experiment contains the most leaves?

.... [1]

iii) Write down **two** differences in the patterns of results shown by the leaves from the two areas.

1. ....

2. .... [2]

c) The students wanted to investigate if the differences between the ivy plants in the two areas were due to the genes of the plants or due to the environment.

Describe how they could investigate this.

....

....

.... [3]

2 This question is about adaptation.

Camels are adapted to living in hot deserts.

a) Table 1 shows some details about humans and camels living in a desert.

Table 1

	Human	Camel
Body mass (kg)	80	400
Mass of sweat made to keep body temperature constant (kg/hour)	1.0	0.3
Amount of sweat that can be lost without dying (% of body mass)	10%	20%

i) How many times greater is the mass of a camel compared to a human?

_____ times   [1]

ii) In the desert, a man needs to lose 24 kg of sweat in 24 hours to keep a constant body temperature.

Calculate the mass of sweat that a camel needs to lose.

_____ kg   [2]

iii) Compare the maximum number of kilograms of sweat that a human and a camel can lose without dying.

_____

_____

_____   [3]

b) Camels also have adaptations to their blood.

Table 2 gives information about the red blood cells in camels and humans.

Table 2

	Number of red blood cells per mm³ of blood	Volume of one red blood cell in arbitrary units
Camel	6	40
Human	5	80

i) Use **Table 2** to put a circle around the words that complete these sentences.

In 1 mm³ of camel blood, there are **fewer / more / the same number of** red blood cells compared to 1 mm³ of human blood.

One red blood cell in a camel is **smaller than / larger than / the same size as** one red blood cell in a human.   [2]

ii) The percentage of the blood that is made up of red blood cells is given by this formula:

$$\% \text{ of blood} = \frac{\text{Number of red blood cells per mm}^3 \text{ of blood} \times \text{Volume of one red blood cell}}{10}$$

In human blood, 40% of the blood is red blood cells.

Calculate the percentage of camel blood that is red blood cells.

_____ %   [2]

iii) If an animal does not drink much water, then the percentage of red blood cells can increase. This can be dangerous.

Explain how camel blood is better adapted to deserts than human blood.

....

....

.... [3]

Total Marks .... / 22

Testing Understanding

1 The drawings and information below are about some extinct animals.

Giant moa
Became extinct 400 years ago.
Was overhunted for meat.

Woolly mammoth
Became extinct in about 6000 BC due to the climate getting warmer.

Tasmanian tiger
Became extinct in 1936.
Was killed to protect farmers' sheep.

Polynesian tree snail
Became extinct in 1996.
Was eaten by another species of snail brought to the islands by humans.

a) Which of these animals was **not** made extinct due to the actions of humans?

.... [1]

b) Write the names of the animals in the order that they became extinct.

Start with the first to become extinct.

.... [1]

c) Snails, like the Polynesian tree snail, can have many different patterns on their shells.

Snails living in different habitats have different patterns.

Draw a line to match each statement about the snails with the correct term.

Statement   Term

| Snails can be born with different patterns of bands |

| Competition |

| Different snails 'fight' each other for food |

| Evolution |

| Over many years the population of snails in a habitat changed |

| Variation |

[2]

2 The pictures show some characteristics of a mother, father and their four children.

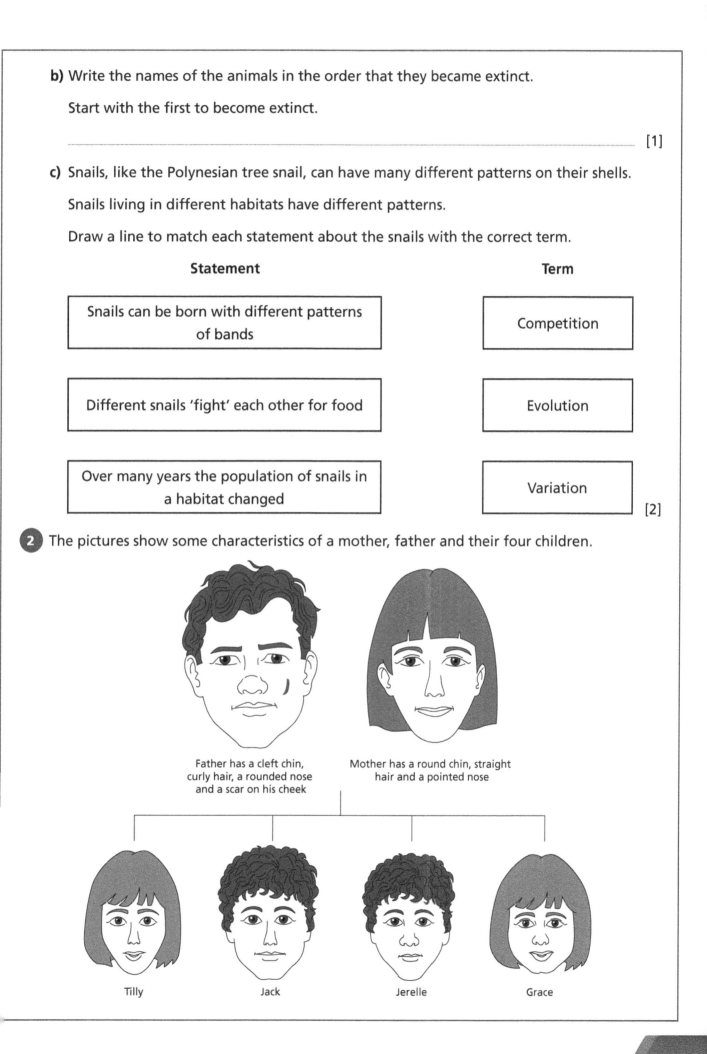

Father has a cleft chin, curly hair, a rounded nose and a scar on his cheek

Mother has a round chin, straight hair and a pointed nose

Tilly   Jack   Jerelle   Grace

a) Which characteristic of the father was determined by the environment only?

_____ [1]

b) Which characteristics did Grace inherit from her father?

_____ [2]

c) The cleft chin shows discontinuous variation.

Explain what this means.

_____

_____ [1]

d) Tilly and Grace are twins.

i) How can you tell that they are not identical twins?

_____ [1]

ii) How are identical twins formed?

_____

_____ [2]

3 The diagram shows the chromosomes from a human cell.

The chromosomes are arranged in pairs.

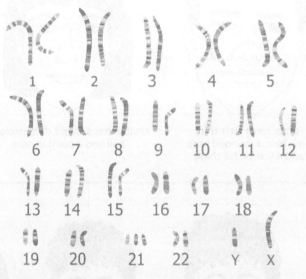

a) Why do people usually have two copies of each chromosome?

_____

_____ [1]

b) How many chromosomes does this person have in every body cell?

.... [1]

c) This person has a genetic condition.

How can you tell this from the diagram?

.... [1]

d) What sex is this person? Explain how you can tell this from the diagram.

....

.... [2]

4 In 1953, Watson and Crick published a paper showing the structure of DNA.

a) Circle the diagram that shows the DNA structure. [1]

A   B   C   D

b) Several other scientists were working on DNA. Their work helped Watson and Crick to find the structure.

How did Rosalind Franklin's work help Watson and Crick? Tick **one** box.

She used X-rays to get data about the size and shape of DNA. ☐

She took photographs of DNA using an electron microscope. ☐

She built a model of DNA to show how the nucleotides fit together. ☐   [1]

c) Erwin Chargaff worked out the percentage of each of the four bases in DNA from different organisms.

The table shows some of these percentages.

Organism	% of each base in DNA			
	A	G	C	T
Octopus	32	18	18	32
Chicken	28	22	22	28
Human	29	21	21	29

i) Describe any pattern shown in the data.

....

.... [2]

ii) Explain how this information helped Watson and Crick.

....

.... [2]

iii) In bacteria, 15% of the bases in DNA are adenine (A).

Calculate the percentage of the other bases in bacterial DNA.

T = ... %

G = ... %

C = ... %   [3]

5 Read this information about squirrels and answer the questions that follow.

What has happened to the red squirrels?

In England there are two species of squirrel – red and grey. The red squirrel has lived here for centuries but the grey squirrel was introduced from America in the 1890s.

In the last 70 years, the number of red squirrels has been going down and they are now rare. Scientists are trying to find out why.

One idea was that the greys passed on a disease to the reds. However, no evidence has been found to support this.

Most of the woods in England are deciduous woodlands. Both types of squirrel eat acorns found in the woodlands but the reds find it hard to digest the acorns. This is because their enzymes are less efficient at digesting acorns. They therefore need to eat more acorns than the grey squirrels to get the same amount of energy.

a) Scientists have concluded that the grey squirrels are outcompeting the red squirrels in English woodlands.

   i) What are the squirrels competing for?

   .... [1]

   ii) Why are the grey squirrels outcompeting the red squirrels?

   .... [1]

b) Scientists hope that the red squirrel might evolve to be able to digest acorns.

   Use ideas about natural selection to explain how this might happen. Include the words from the box in your answer.

adapted	reproduce	mutation	survive

   ....

   ....

   ....

   ....

   .... [4]

Total Marks .... / 31

P14, Q1a)-b)

Working Scientifically

1 Members of the same snail species can have different shell patterns – banded and unbanded.

A scientist plans an investigation to see how long different snails stay in the sunlight. They used a special paint to mark the snail shells. The paint fades when exposed to daylight.

This is the scientist's method:

1. Collect 100 snails with banded shells and 100 snails with unbanded shells.
2. Put a small spot of paint on the top of each of the snails' shells.
3. Put the snails into cages planted with grass and nettles.
4. After 60 days, measure how much the paint has faded.

a) Why did the scientist put the spot of paint in the same place on each snail?

_____

_____ [1]

b) In this experiment, the scientist did not have to worry about the paint spot making the snails easier to spot by bird predators.

Why is this?

_____ [1]

c) These are the scientist's results. The higher the number, the more faded the paint.

	Banded snails	Unbanded snails
Average fading of paint spot on shell	4.3	3.9

i) Explain how the scientist compared how long the two types of snails stayed in the sunlight.

_____

_____

_____ [2]

ii) Explain the scientist's results. In your answer use the words from the box below.

camouflage	sunlight	predators

_____

_____

_____ [3]

2 Scientists have studied a lizard called the brown anole that lives on islands in the Caribbean Sea. They set up an experiment to see if the lizard changed in response to predators.

This is the scientists' method:

    1. Measure the length of the legs of the lizards on 12 different islands.
    2. Introduce a new predator of the brown anole on six of the islands.
    3. Keep the six other islands free of the predator.
    4. Re-measure the length of the legs of the lizards after 6 months and 12 months.

a) The scientists left six islands free of the predator.

Explain why they did this.

....

.... **[2]**

b) Suggest what happened to the number of lizards on the different islands over the 12 months.

Explain your answer.

....

....

.... **[3]**

c) The scientists made these observations about the lizards on the six islands with the newly introduced predator:

    • At first, the lizards seemed to try to run away to escape the predator.

    • After six months, the average leg length of the brown anoles had increased.

  i) Use Darwin's theory of natural selection to explain how the average leg length of the lizards increased.

....

....

....

.... **[4]**

ii) Over the next six months, the anoles on the islands with the new predator started to spend more time in the trees. The scientists found that the average leg length was now shorter.

Suggest why their legs were shorter.

....

....

....
[2]

d) The scientists think that this experiment gives evidence for natural selection happening.

Why is it difficult for scientists to design experiments that show natural selection happening?

....
[1]

Total Marks .... / 19

Science in Use

1 Read the passage and then answer the questions that follow.

Vitamin D

Approximately 1 in 6 people in Britain have low levels of vitamin D, which can cause a deficiency disease. Our main source of vitamin D comes from the action of sunlight on the skin. Sunlight converts provitamin D into an active form of vitamin D that our bodies need. However, in the UK there is only enough sunlight between April and September. This means we must rely on dietary sources of vitamin D for the rest of the year.

Scientists have now created genetically edited tomatoes, each containing as much vitamin D as two eggs. The tomato plants were made by making small changes to a tomato gene. The gene codes for an enzyme that normally converts provitamin D into cholesterol. By altering this enzyme, provitamin D builds up in the tomatoes. Then by growing the fruit outdoors, the provitamin D would be converted to vitamin D by sunlight.

Gene editing (GE) is a new technology. The older technique of genetic modification (GM) involves putting genes into an organism, sometimes from a completely different species. Any new GM or GE crop must undergo testing, which can take around five years. However, the UK government wants GE to be able to be used more quickly.

a) What is a deficiency disease?

....
[1]

b) Which is a symptom of a lack of vitamin D? Tick the correct box.

bleeding gums ☐

lack of red blood cells ☐

soft bones ☐ [1]

c) The population of the UK is about 70 million.

Calculate how many people have low levels of vitamin D in the UK.

.... million [2]

d) Explain why changing one gene in tomatoes can cause provitamin D to build up.

....

.... [2]

e) Suggest **one** reason why GE or GM crops need to be tested before they are allowed to be grown.

.... [1]

f) Scientists think that GE crops need less testing than GM crops. Suggest a reason why.

....

.... [1]

2 Read the article and then answer the questions that follow.

Muscular dystrophy is a disorder that occurs in about 0.03% of boys that are born. It is a genetic condition that causes the muscles of the body to waste away.

It is caused by an error in a gene on the X chromosome. Boys only have one X chromosome, so they will have the disorder if this chromosome is affected. The gene codes for a chemical called dystrophin. This chemical is missing in people with muscular dystrophy.

People with the condition usually die in their 20s because of weakness in their heart and lung muscles. At the moment, there is no cure.

Women who have one faulty X chromosome have a 50% chance of passing it on to their sons. It is now possible for a woman to have a test as early as seven weeks into their pregnancy to see if they are going to have a boy or a girl. This test can be done much earlier than having a scan. The test works by looking for small parts of the baby's Y chromosome that entered the woman's blood.

The test, however, can leave the woman/parent(s) with a difficult decision to make.

a) Each year about 400 000 boys are born in the UK.

Calculate how many of them would have muscular dystrophy.

[2]

b) Which of these is a reason why very few girls have muscular dystrophy? Tick the correct box.

Females have two copies of the X chromosome. ☐

Females do not need dystrophin. ☐

Females inherit a Y chromosome from their father. ☐ [1]

c) What type of chemical is dystrophin? Tick the correct box.

DNA ☐

fat ☐

protein ☐ [1]

d) Explain why the presence of parts of the Y chromosome in women's blood shows that they are pregnant with a boy.

[1]

e) Discuss why the development of this test might leave the pregnant woman/parent(s) with a difficult decision to make.

[2]

| Total Marks | / 15 |

	Vocabulary Builder	Maths Skills	Testing Understanding	Working Scientifically	Science in Use
Total Marks	/ 20	/ 22	/ 31	/ 19	/ 15

Vocabulary Builder

1 The passage describes the effects of different types of drugs.

Fill in the spaces using words from the box.

alcohol	aspirin	cocaine	depressants	hallucinogens
	LSD	nervous	stimulants	respiratory

Some drugs are taken because they block pain messages in the _____ system.

An example of a painkiller that is taken legally is _____.

Drugs that increase brain activity and alertness are called _____.

An example of one of these drugs is _____.

Other drugs, such as _____ will slow down activity in the brain.

They are called _____.   [6]

2 Draw lines to join each drug to its correct category.

Drug	Category
Codeine	Depressant
Caffeine	Hallucinogen
Tranquillisers	Painkiller
LSD	Stimulant

[3]

3 This question is about different drugs. For each statement put a tick to show if the statement is **true** or **false**.

a) Morphine

	True	False
Is only legally available with a prescription	☐	☐
Acts as a stimulant	☐	☐
Is a strong painkiller	☐	☐

[3]

b) **Tobacco**

	True	False	
Causes cirrhosis of the liver	☐	☐	
Only affects the person who is smoking the drug	☐	☐	
Significantly increases the risk of developing lung cancer	☐	☐	[3]

c) **Anabolic steroids**

	True	False	
Can increase muscle growth	☐	☐	
Increase levels of natural hormones in the body	☐	☐	
Can cause liver damage	☐	☐	[3]

Total Marks _____ / 18

Maths Skills

1 Cans of beer are labelled with the number of units of alcohol that they contain.

One unit of alcohol is 10 millilitres of pure alcohol.

The table gives data about four different brands of beer.

Brand of beer	Volume of beer in a can (ml)	Units of alcohol in a can
P	330	1.8
Q	440	1.7
R	330	1.5
S	275	1.0

a) Calculate the volume of pure alcohol in one can of brand **Q**.

_____ ml [1]

b) Calculate the percentage of alcohol in one can of brand **Q**.

Give your answer to 2 significant figures.

_____ % [3]

c) It takes one hour for the liver of an average adult to remove one unit of alcohol from the blood.

If an average adult drinks two cans of brand **P**, calculate how many minutes it would take to remove all the alcohol from their blood.

.... minutes [2]

d) The graph shows how different levels of alcohol in the blood can affect the chance of a person having an accident when driving a car.

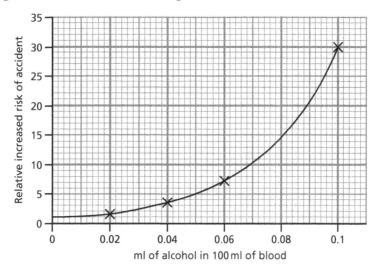

A person drinks four cans of brand **S**. For each unit of alcohol drunk, the concentration of alcohol in the blood increases by 0.02 ml/100 ml of blood.

i) Calculate the concentration of alcohol in the person's blood straight after they have drunk three cans of beer.

.... ml of alcohol in 100 ml of blood [2]

ii) Use the graph to find the increased risk of the person having an accident if they drove a car after drinking the beer.

.... [1]

Total Marks .... / 9

Testing Understanding

1 George and Lucas are talking about the illegal drug cannabis.

> **George**
>
> I think cannabis should only be a class C drug. It is only mildly hallucinogenic.

> **Lucas**
>
> I think that cannabis should stay as a class B drug. It slows people's reactions and can increase the chance of schizophrenia.

a) Illegal drugs are put into different classes – A, B or C.

Suggest why this is.

_____

_____

_____ [2]

b) George says that cannabis is mildly hallucinogenic.

What does **hallucinogenic** mean?

_____

_____ [1]

c) Lucas says that cannabis slows people's reactions.

Explain why this could be dangerous.

_____

_____ [2]

d) Studies have shown that daily users of cannabis have a 2% chance of developing schizophrenia. This is compared to a 1% chance in non-users.

Discuss whether this shows that cannabis causes schizophrenia.

_____

_____

_____ [2]

2 In 2010, a group of scientists tried to decide which drugs are most harmful. They gave each drug a score based on how harmful it is to the person using the drug and to other people in society.

The results for the top ten scoring drugs are shown in the graph.

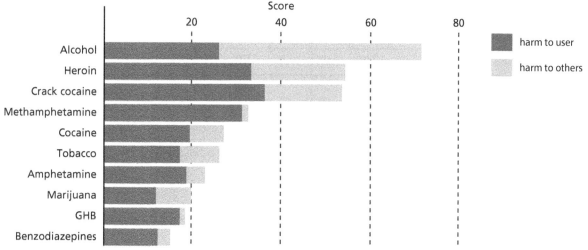

SOURCE: Independent Scientific Committee on Drugs, based on analysis of UK drug use, *The Lancet*, 2010

a) Which drug was highest scoring for harm to users?

.... [1]

b) Which drug was the highest scoring for harm to others?

.... [1]

c) Which two drugs caused the least harm to others?

.... [2]

d) Explain how using tobacco and alcohol can cause harm to others.

Tobacco: ....

Alcohol: .... [2]

3 Scientists can compare how dangerous drugs are by working out a figure called the 'therapeutic ratio'.

This ratio is worked out using this equation:

$$\text{therapeutic ratio} = \frac{\text{dose that could kill}}{\text{smallest dose that has an effect}}$$

The table gives information about three drugs.

Drug	Dose that could kill (mg)	Smallest dose that has an effect (mg)	Therapeutic ratio
Alcohol	300 000	30 000	10
Cocaine	1200	50	
Heroin	50	10	5

a) Calculate the therapeutic ratio for cocaine.

Show your working here and then write your answer in the table. [2]

b) Explain why heroin is thought to be the most dangerous drug of the three in the table.

Use the data in the table to explain your answer.

_____

_____

_____ [2]

Total Marks _____ / 17

Working Scientifically

1 For many years, scientists thought that there was a connection between tobacco and cancer.

In 1751, Dr John Hill noticed that six of his patients who used tobacco also had cancer of the nose.

In 1950, Dr Richard Doll studied thousands of patients in London hospitals. The table shows Doll's data.

	Number of deaths in a year per thousand patients	
	Non-smokers	Heavy smokers
Deaths due to lung cancer	0.0	1.1
Deaths by all causes	13.6	16.3

a) A hospital had 3000 non-smoking patients a year.

Calculate how many of those patients would be expected to die in a year.

_____ [2]

b) Put a tick next to any conclusions that can be made from Doll's data.

Non-smokers are very unlikely to get lung cancer. ☐

For every 2000 heavy smokers, over 34 would die of lung cancer. ☐

Non-smokers have the same death rate as heavy smokers. ☐

Compared to non-smokers, heavy smokers are more likely
to die from other diseases as well as from lung cancer. ☐ [2]

c) Give a reason why Doll's data provides better evidence for the connection between tobacco and cancer than Hill's data.

.... [1]

d) Most people agreed that Doll's data showed a correlation between smoking and cancer.

Explain what the term **correlation** means.

....

.... [2]

e) In 1965, Austin Bradford Hill suggested different features of a study that would help to show that a factor causes an effect and doesn't just show a correlation.

Here are three of his suggested features:

    1. The same results should be found in different places.
    2. The larger the factor, the greater the effects.
    3. The factor must happen before the effect.

Explain how Richard Doll could change his study to include these three features.

Feature 1: ....

....

Feature 2: ....

....

Feature 3: ....

....

[3]

Total Marks .... / 10

Science in Use

1 Read the passage about marijuana and answer the questions that follow.

Is marijuana safe as a medicine?

Marijuana is the dried leaves and flowers of the *Cannabis* plant. About two thousand years ago people grew *Cannabis* to make clothing and rope. People then started to realise that eating or smoking marijuana could produce effects on the brain. It contains a chemical called THC which distorts how the brain perceives the world. Marijuana initially had low levels of THC, but as people started to eat or smoke marijuana for its effects, new strains were grown with much higher concentrations of THC.

Using marijuana for long periods has been linked to depression and anxiety and possibly schizophrenia. It is also addictive. It has therefore been made an illegal drug in the UK.

More recently it has been found that marijuana can be used to help people who have certain types of epilepsy. It is also used for people with multiple sclerosis, which causes muscle stiffness and spasms. The liquid used for epilepsy is called Epidyolex and does not contain THC. The drug used by people with multiple sclerosis is called Sativex and contains a chemical called CBD and THC. In the UK, both treatments are only available on prescription and only when other treatments have not worked.

Some people say that marijuana or *Cannabis* extracts could help people suffering from other conditions, but it has been difficult to find evidence for this.

a) Marijuana is an addictive drug.

What does **addictive** mean?

_____

_____ [2]

b) Different strains of *Cannabis* have very different concentrations of THC.

Explain why this can cause problems for people who use marijuana illegally.

_____

_____ [2]

c) Explain why Sativex is likely to produce more side effects than Epidyolex.

....

.... [2]

d) Suggest why Sativex is only available with a prescription.

....

.... [1]

e) There are few scientific studies on the effects of marijuana on various medical conditions. Suggest reasons why.

....

.... [2]

Total Marks .... / 9

	Vocabulary Builder	Maths Skills	Testing Understanding	Working Scientifically	Science in Use
Total Marks	/ 18	/ 9	/ 17	/ 10	/ 9

Vocabulary Builder

1 Match the type of material to its description.

Key word	Definition
Polymer	Material made from two parts – binder (matrix) and reinforcement
Ceramic	Very large molecules made from small repeating units
Refractory ceramic	Inorganic, non-metallic solid that retains its strength at high temperatures
Composite	Inorganic, non-metallic solid

[3]

2 Complete the sentences about energy change in reactions using the words from the box. You can use each word once, more than once or not at all.

making	breaking	release	absorb	decomposition	combustion

Endothermic reactions .... energy from the surroundings and cause the temperature to drop. .... bonds and chemical reactions like in .... are examples of endothermic changes.

Exothermic reactions .... energy to the surroundings and cause a temperature rise. .... bonds and chemical reactions like .... are examples of exothermic changes.

[6]

3 For each of the following chemical equations, classify the type of chemical reaction happening. Choose from the words in the box below. You can use each word once, more than once or not at all.

smelting	decomposition	displacement	oxidation

a) calcium carbonate → calcium oxide + carbon dioxide .... [1]

b) magnesium + iron sulfate → magnesium sulfate + iron .... [1]

c) copper oxide + carbon → copper + carbon dioxide .... [1]

d) sodium + oxygen → sodium oxide .... [1]

e) aluminium + iron oxide → iron + aluminium oxide .... [1]

f) copper carbonate → copper oxide + carbon dioxide .... [1]

4 What is the name of a rock that contains a sufficient quantity of the metal to make it economically worth extracting the metal?

Tick **one** box. [1]

mineral	☐	native	☐
ore	☐	alloy	☐

5 Which of the following statements about **catalysts** are **true** and which are **false**?

Write **true** or **false** in the spaces provided.

a) Catalysts are a reactant in a chemical reaction. .... [1]

b) Catalysts can only speed up chemical changes. .... [1]

c) Enzymes are biological catalysts. .... [1]

d) Most catalysts give a lower activation energy and an alternative

reaction pathway. .... [1]

e) Catalysts in a catalytic converter are not part of the chemical reaction. .... [1]

Total Marks .... / 21

Maths Skills

1 The Earth's crust has been analysed to find out the elements that are most commonly found there.

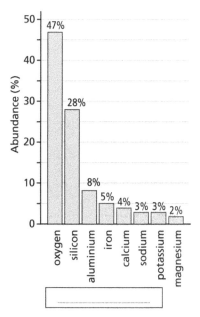

a) How has the data been presented?

.... [1]

b) Complete the *x*-axis label. [1]

c) What is the most abundant element in the Earth's crust?

.... [1]

d) What percentage of the Earth's crust is made of metals?
Show your working.

.... [2]

2 Hydrogen peroxide can decompose into water and oxygen. The rate of this reaction can be monitored by measuring the volume of oxygen made. The data that is collected can be used to draw a graph.

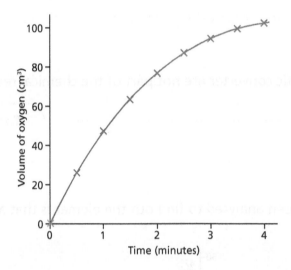

a) What is the dependent variable? .... [1]

b) What is the unit of the independent variable? .... [1]

c) How much gas was collected at 4 minutes? Give the unit in your answer. .... [1]

d) Use the graph to describe the rate of this reaction.

....

....

....

[3]

3 Jo was completing an experiment to make pure copper from the smelting of copper oxide. She carefully measured 1.60 g of copper oxide and calculated that she could make a maximum of 1.27 g copper metal. When she completed the smelting experiment, she actually only collected 0.9 g.

Calculate the percentage yield for Jo's experiment. Give your answer to 1 decimal place.

yield = ... [4]

4 Metals were discovered at different times in history as technology and scientific knowledge developed. The less reactive metals like gold were found sooner than the reactive metals like aluminium. A timeline of the discoveries of key metals can be drawn.

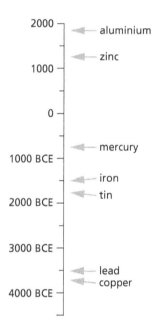

a) Which metals were discovered more than 4000 years ago?

.... [2]

b) Approximately how many years were there from the discovery of zinc to the discovery of aluminium? .... [1]

c) In approximately what year was mercury discovered? .... [2]

Total Marks .... / 20

Testing Understanding

1 Ceramics are an important classification of materials. Match the ceramic with its use and the property which makes it suitable for this use.

Key word	Use	Property
Refractory ceramics	Bath tubs	Hard and resistant to wear, chemically stable and non-toxic
Earthenware and stoneware	Lining furnaces	Transparent, durable and a thermal insulator
Porcelain	Plates, dishes and cups	Very hard and durable, chemically stable and non-toxic
Glass	Window panes	Retains strength at high temperatures

[4]

2 Part of the reactivity series is shown below.

Potassium
Sodium
Calcium
Magnesium
Aluminium
Zinc
Iron
Copper
Gold

a) What is the reactivity series?

_____ [2]

b) What is a displacement reaction?

_____ [1]

c) Complete the following word equations for displacement reactions:

    i) zinc chloride + calcium → .... + ... [2]

    ii) copper sulfate + magnesium → .... + ... [2]

    iii) magnesium nitrate + copper → .... [1]

d) Why could gold not undergo any displacement reactions with the metals in this reactivity series list?

.... [1]

e) Why is carbon often added to the reactivity series?

.... [2]

3 Polymers are found in all living organisms and are made from very long molecules.

a) What are the small repeating units called that make up a polymer?

.... [1]

b) Glucose is made by plants during photosynthesis.

What is the name of the polymer that plants use to store the extra sugar that they have made?

.... [1]

c) Photosynthesis is controlled by enzymes.

What is the name of the polymer that makes up an enzyme?

.... [1]

d) Rubber is a natural polymer made from the sap of the rubber tree.

Why is rubber strong and elastic?

....

....

.... [3]

4 Complete the sentences about synthetic polymers using the words from the box. You can use the words once, more than once or not at all. [7]

crude oil	polymer	higher	lower	ores
monomer	polyethene	catalyst	cross links	

Synthetic polymers are usually made by heating a monomer under high pressure with a ....

Different monomers make different polymers. For example, ethene makes .... and polystyrene is made from the styrene .... Most monomers are made from extracts of ....

The structure of the .... affects its physical properties. The longer the polymer chain and the more branched it is, the .... the melting point. Polymers that have .... between chains have a high strength and a high melting point.

5 Iron is extracted from haematite in a blast furnace. This reaction is exothermic.

a) What type of chemical equation happens in the blast furnace?

.... [1]

b) Complete the word equation for the production of iron in the blast furnace.

iron oxide + carbon → .... + ... [2]

c) Explain why carbon can be used to extract iron from its ore.

....

.... [2]

d) Draw an energy level diagram for the reaction in the blast furnace.

[4]

6 Many metals are extracted from the Earth in mines. A mine is only worked for up to 70 years and then is abandoned.

a) How can toxic materials from the waste rock of the mine pollute water?

....

.... [2]

b) Some mining companies use reclamation to restore the landscape.

What is **reclamation**?

_____

_____ [2]

c) How can recycling reduce the environmental impact of mining?

_____

_____

_____

_____ [4]

Total Marks _____ / 45

Working Scientifically

1 Kareem wanted to investigate the reduction of copper oxide.

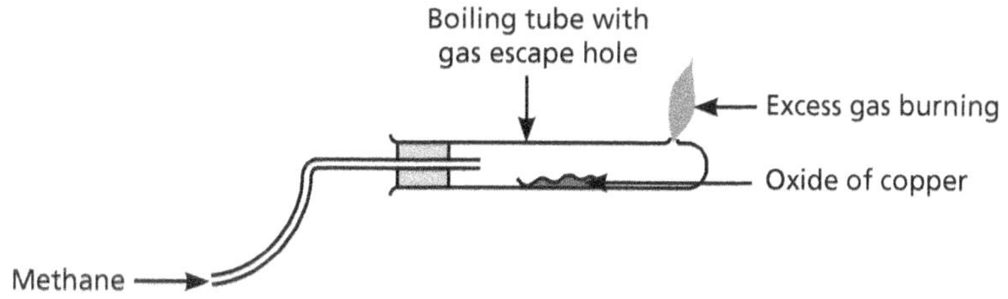

He carefully measured the mass of different equipment and substances. He recorded the results in a table.

Equipment	Mass (g)
Boiling tube	20.45
Boiling tube + copper oxide	23.65
Boiling tube at the end of the experiment	23.01

a) What piece of equipment is missing from the diagram that Kareem would need to use to heat the apparatus?

_____ [1]

b) Where would Kareem get methane from in a laboratory?

_____ [1]

c) What piece of equipment would Kareem use to measure mass?

.... [1]

d) What is the mass of copper oxide that Kareem used in his experiment?

....g [2]

2 Priya wanted to investigate displacement reactions to find out if they were exothermic or endothermic. She carefully measured 20 cm³ of copper sulfate solution into a test tube and added 0.1 g of magnesium and observed the reaction.

a) What piece of equipment would Priya use to measure the copper sulfate solution?

.... [1]

b) What is the dependent variable in this investigation?

.... [1]

c) Describe how Priya could observe the experiment to get valid results.

....

.... [2]

d) Explain how the results can be used for Priya to make a conclusion.

....

.... [2]

3 Eddie decided to investigate the reactivity of different metals to try to make a mini reactivity series. He decided to compare the reaction of the metals in acid. He chose to use sodium, zinc and copper.

a) What is the independent variable in Eddie's experiment?

.... [1]

b) List **three** control variables for Eddie's experiment.

....

....

.... [3]

c) Why is Eddie's experiment not safe?

....

.... [1]

d) Suggest how Eddie could complete this investigation in a safer way.

....

.... [2]

Total Marks .... / 18

Science in Use

1 Read the passage below about concrete and then answer the questions that follow.

The Romans were the first people to make concrete about 2000 years ago. The Flavian Amphitheatre, more commonly known as the Colosseum, is an example of a Roman building that is still standing today!

Roman concrete, like modern concrete, is a composite material made from small stones called aggregate held together by cement. *Opus caementicium* (otherwise known as Roman concrete) is much more durable than modern concrete as the cement chemically changes and makes a stone-like substance. Modern cement breaks down after a few decades and even quicker if it gets wet.

a) Define the term **aggregate**.

.... [1]

b) Give **one** use of concrete.

.... [1]

c) Describe a composite.

[2]

d) Justify why concrete is classified as a composite.

[3]

2 Read the passage below about the production of steel and then answer the questions that follow.

Steel is an alloy of mainly iron. Iron can be extracted from its ore called haematite, which is mainly iron oxide. The ore is put into a blast furnace with coke and limestone. Iron oxide is reduced to make iron, and the acidic impurities in the process react with the limestone to make a substance called slag. Slag can be used for road building and breeze blocks.

a) What compound in haematite contains the iron?

[1]

b) What are the **two** products from the blast furnace that can be used in construction?

[2]

c) Carbon in the coke reacts with the iron ore to make carbon dioxide and iron. Write a word equation for this reaction.

[3]

d) Give **one** environmental problem of the waste gases produced by the blast furnace.

[1]

Total Marks .... / 14

	Vocabulary Builder	Maths Skills	Testing Understanding	Working Scientifically	Science in Use
Total Marks	/ 21	/ 20	/ 45	/ 18	/ 14

Vocabulary Builder

1. Which of the following statements about **sustainability** are **true** and which are **false**?

   Write **true** or **false** in the spaces provided.

   a) Fossil fuels are a renewable resource. .... [1]

   b) Down-cycled products can be recycled. .... [1]

   c) Recycling is more energy efficient than getting new materials. .... [1]

   d) When humans process natural resources, pollution is produced. .... [1]

   e) Biological resources like plants and animals are renewable resources. .... [1]

2. Match the type of rock to its description.

Rock	Definition
Extrusive igneous	Formed from magma below the Earth's surface in the crust
Metamorphic	Rock formed by compressing small fragments of rock
Sedimentary	Formed when magma flows onto the Earth's surface
Intrusive igneous	Rock formed when other rocks are heated and put under a lot of pressure

   [3]

3. Complete the sentences about energy change in reactions using the terms from the box. You can use each term once, more than once or not at all.

carbon	carbon dioxide	methane	cycle
footprint	greenhouse	atoms	molecules

   Carbon .... are constantly moving into and out of organisms, the atmosphere and the Earth in the carbon cycle. Humans are changing the balance of the carbon ....
   and this can lead to environmental problems. So, it can be useful to calculate the carbon
   .... to help measure the impact on the environment.

The carbon footprint calculates the total amount of .... gases that you are expected to make in units of .... The world average is about 4 tonnes of .... per person per year, but in developed countries, like the UK, it is often higher.

[6]

4 Air pollution is when gases are released into the air that may not be naturally present in the atmosphere or are released in higher quantities than are found naturally. For each of the following environmental problems, classify which gas(es) in the atmosphere causes the problem. Choose from the terms in the box below. You can use each term once, more than once or not at all.

methane	carbon dioxide	sulfur oxide	nitrogen oxide	CFCs

a) Reduction in the ozone layer   ....   [2]

b) Global warming   ....   [2]

c) Acid rain   ....   [2]

d) Smog   ....   [1]

Total Marks .... / 21

Maths Skills

1 Air is a mixture of gases. The pie chart below shows the composition of dry air.

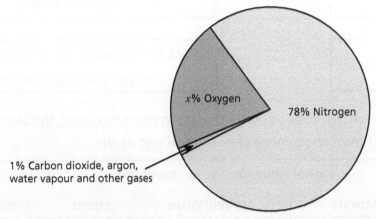

x% Oxygen

78% Nitrogen

1% Carbon dioxide, argon, water vapour and other gases

a) What percentage of dry air is oxygen?   .... % [1]

b) What is the approximate ratio of oxygen : nitrogen in dry air?

....   [2]

c) A scientist collected a 120 g sample of dry air.

What mass of the sample would be nitrogen? Show your working and give your answer to 2 significant figures.

.... g [3]

2 When products come to the end of their useful life, they are disposed of. The table shows some data about the time it takes for substances to completely decompose.

Substance	Time (weeks)
Paper	2–4
Paper milk cartons	260
Plastic bags	53 040
Metal drinks cans	5200
Glass bottles	26 000

a) What is the unit of the dependent variable in this investigation? .... [1]

b) How many more weeks does a glass milk bottle take to decompose compared to a paper milk carton?

.... weeks [2]

c) How many years does it take for a plastic bag to decompose?

.... years [2]

d) What is the least number of minutes that it takes for paper to decompose? Give your answer in standard form.

.... s [5]

3 Carbon dioxide is a greenhouse gas and is carefully monitored in our atmosphere. The graph below shows how the percentage of carbon dioxide has changed since 1700.

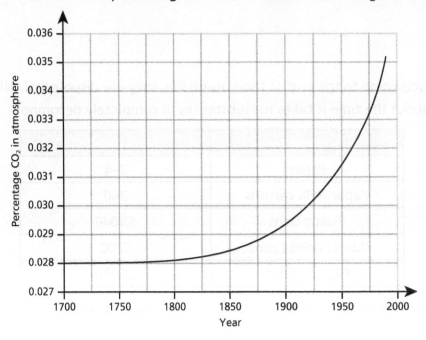

a) What was the percentage of carbon dioxide in the air in each of the following years?

   i) 1700 _____ % [1]

   ii) 1750 _____ % [1]

   iii) 1950 _____ % [1]

b) Approximately how much more carbon dioxide was there in the air in 1975 than in 1800?

   _____

   _____ [3]

c) In a 100 g sample of the atmosphere, what mass of the sample would **not** be carbon dioxide in 1700? Give your answer to the nearest whole percentage.

   _____

   _____

   _____ [3]

Total Marks _____ / 25

1 Rocks are the solid material that makes the Earth's surface. There are three categories of rocks determined by how they formed. Put a tick in the table to show which features are found in each type of rock.

Rock type / Feature	Igneous	Metamorphic	Sedimentary
Contains layers			
Most likely to have fossils			
Contains particles like sand and pebbles			
Contains crystals			

[6]

2 The structure of the Earth has different layers. Use the words in the box to answer the questions that follow.

Inner core	Mantle	Crust	Outer core

a) Label the diagram of the Earth.

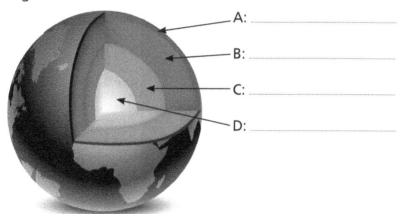

A: ....

B: ....

C: ....

D: ....

[3]

b) Which parts of the Earth are liquid?

.... [2]

c) What **two** metallic elements are found in the Earth's core?

.... [2]

d) Which **three** oxides are found in the mantle?

....

.... [3]

e) Explain how magma is related to lava.

....

.... [2]

3 The carbon cycle can be represented in the diagram below.

Carbon dioxide CO_2
Atmosphere

A

B

B

C

Plants

Animals

Factories
Power stations
Vehicle emissions

D

Dead organic
matter

E

Extraction and
transportation

Which letter in the diagram represents the following?

a) Respiration .... [1]

b) Photosynthesis .... [1]

c) Fossil fuels, e.g. oil, coal and natural gas .... [1]

d) Combustion of fossil fuels .... [1]

e) Decay of organic matter .... [1]

4 Milk bottles are often made from glass, or PET, which is a type of plastic.

a) Describe how glass milk bottles can be re-used.

....

.... [3]

b) Describe how both glass and plastic bottles can be recycled.

....

.... [3]

c) Evaluate why glass bottles are less likely to be down-cycled than plastic milk bottles.

....

....

....

.... [4]

5 Which of the following statements about the development of the Earth are **true** and which are **false**?

Write **true** or **false** in the spaces provided.

a) The Earth is about 4.6 million years old. .... [1]

b) The composition of the early Earth's atmosphere is thought to be similar to the composition of the modern-day atmospheres of Mars and Venus. .... [1]

c) Oxygen in the atmosphere is made by photosynthesis. .... [1]

d) The Earth has warmed since it has formed. .... [1]

e) The oceans are condensed water from the atmosphere. .... [1]

f) Carbon dioxide from the atmosphere dissolved in oceans and was taken in by plants and algae for photosynthesis. .... [1]

Total Marks .... / 39

Working Scientifically

1 Daphne decided that she wanted to investigate how temperature affects crystal size. She used a harmful substance, called salol, which melts at 41.5°C. She melted it and then put some of the melted substance on a cool microscope slide and some on a warm microscope slide. She then waited for the substance to crystalise.

a) What is the independent variable in this experiment?

.... [1]

b) What safety precautions should Daphne take during her experiment?

[2]

c) Describe the observations that Daphne could make.

[3]

d) Daphne's experiment is used to model crystallisation that happens in some rocks.

What type of rock is Daphne modelling in her experiment?

[1]

2 Acid rain is formed when pollutant gases dissolve in rainwater and lower its pH. Some rocks can react with acid and this is a type of chemical weathering.

A student wanted to investigate the reactions of different rocks with acid. He put each of the rock samples into 25 cm³ of acid and observed what happened. He used the equipment below to monitor the reactions:

Cotton wool to stop acid 'spray' escaping

Acid

Rock sample

A

a) What is the name of the equipment labelled as A?

[1]

b) Suggest a suitable acid for this experiment. Justify your choice.

[2]

c) Give **two** control variables for this experiment.

.... [2]

d) If the rock sample reacted, what would the student observe?

....

....

.... [3]

Total Marks .... / 15

Science in Use

1 Read the passage below about air pollution and then answer the questions that follow.

Humans use fossil fuels to power cars, generate electricity, heat our homes and cook our food. Fossil fuels are a mixture of hydrocarbon compounds that are made of only carbon and hydrogen.

As they are made from dead organisms that lived thousands of years ago, they contain impurities like sulfur. When the fuel is combusted, these other elements also oxidise and make acidic gases like sulfur dioxide.

a) What **two** elements are found in hydrocarbon substances?

.... [1]

b) Give an example of a fossil fuel.

.... [1]

c) Write a word equation for the oxidation of sulfur when it is combusted with a fossil fuel.

.... [3]

d) What environmental problem is caused by pollutant acidic gases?

.... [1]

2 Read the passage below about rocks and then answer the questions that follow.

> Limestone and marble are used as building materials. They are both examples of rocks that are mainly made of the compound calcium carbonate. Carbonite is a rare igneous rock that looks very much like marble because it contains at least half dolomite, which is a crystalline form of calcium carbonate found in both of these rocks.
>
> Limestone and chalk are made from the shells of dead sea creatures. These rocks can be exposed to heat and pressure, which causes them to change and make marble.

a) Name **two** sedimentary rocks containing calcium carbonate.

.... [2]

b) Describe how limestone or chalk can be made.

....

....

....

.... [4]

c) Give the name of an igneous rock containing calcium carbonate.

.... [1]

d) Give the name of the mineral found in both carbonite and marble.

.... [1]

Total Marks .... / 14

	Vocabulary Builder	Maths Skills	Testing Understanding	Working Scientifically	Science in Use
Total Marks	.... / 21	.... / 25	.... / 39	.... / 15	.... / 14

Vocabulary Builder

1 Complete the sentences using the words from the box below.

Moon	Earth	galaxy	Sun

a) The Moon is in orbit around the .... [1]

b) The Earth is in orbit around the .... [1]

c) The Sun is in orbit around the centre of the .... [1]

2 A box is stationary on a table. There are two forces acting on the box.

a) One of the forces is an upward contact force exerted by the table on the box.

Give the name and direction of the other force acting on the box.

.... [2]

b) The box is in equilibrium. Explain what this means in terms of the two forces acting on the box.

.... [1]

3 Consider the moving objects listed in the table.

Object	Motion	Location
car	accelerating	motorway slip road
cyclist	constant speed	flat straight road
bus	decelerating	flat straight road approaching traffic lights

Identify which of the objects are in equilibrium and which are not in equilibrium. Explain your answers.

....

....

....

.... [6]

4 Which unit is most suitable for distance measurements to other galaxies? Tick **one** box. [1]

metre ☐   kilometre ☐   light-year ☐   mile ☐

5 The planet Jupiter has more than 70 moons.

What force keeps these moons in orbit around Jupiter?

.... [1]

6 Which statement about the Earth is incorrect? Tick **one** box.

The angle between the Earth's axis of rotation and its axis of orbit is about 23°. ☐

The Earth rotates on its axis from east to west. ☐

It is summer in the UK when the northern hemisphere is tilted towards the Sun. ☐ [1]

Total Marks .... / 15

Maths Skills

1 The Earth's gravitational field strength is 10 N/kg at its surface. This means that the force of gravity (also known as weight) on a mass of 1 kg is 10 newtons.

a) Calculate the weight of a man of mass 80 kg at the surface of the Earth.

.... N [1]

b) i) You can use these symbols to represent quantities:

mass = m, weight = W, gravitational field strength = g

The relationship between these quantities is shown by the formula:

$$W = mg$$

Which is the correct rearrangement of this formula to make m the subject?
Tick **one** box.   [1]

$m = \dfrac{W}{g}$ ☐ $m = Wg$ ☐ $m = \dfrac{g}{W}$ ☐

ii) The weight of a hammer on Earth is measured as 4 N using a newton-meter.

Calculate the mass of the hammer in kilograms.

.... kg [1]

iii) The gravitational field strength on the Moon is 1.6 N/kg.

Calculate the weight of the hammer if it were on the Moon.

.... N [1]

2 Speed, distance and time are related by the formula:

$$\text{speed} = \text{distance} \div \text{time}$$

a) A car travels a distance of 1.5 km in 2.5 minutes.

   i) What is the distance travelled in metres? .... [1]

   ii) What is the time taken in seconds?

.... [1]

   iii) Calculate the average speed in m/s.

.... [1]

b) The formula can be written: **speed** $= \dfrac{\text{distance}}{\text{time}}$

   i) Which is the correct rearrangement of this formula to make 'time' the subject?

   Tick **one** box.

   time $= \dfrac{\text{speed}}{\text{distance}}$ ☐

   time = speed × distance ☐

   time $= \dfrac{\text{distance}}{\text{speed}}$ ☐ [1]

   ii) Calculate how much time it would take for a hiker to walk 8.0 km if their average speed is 3.2 m/s.

   Give your answer in both seconds and in minutes.

.... seconds [1]

.... minutes [1]

3 An athlete is videoed as she runs a 100-metre race.

The video is analysed and the distance she has travelled from the start is recorded every second in the table below.

Distance (metres)	0	5	12	23	34	45	55	66	77	88	95	98	100
Time (seconds)	0	1	2	3	4	5	6	7	8	9	10	11	12

a) Plot the data in the table on the graph paper. [4]

Graph 1

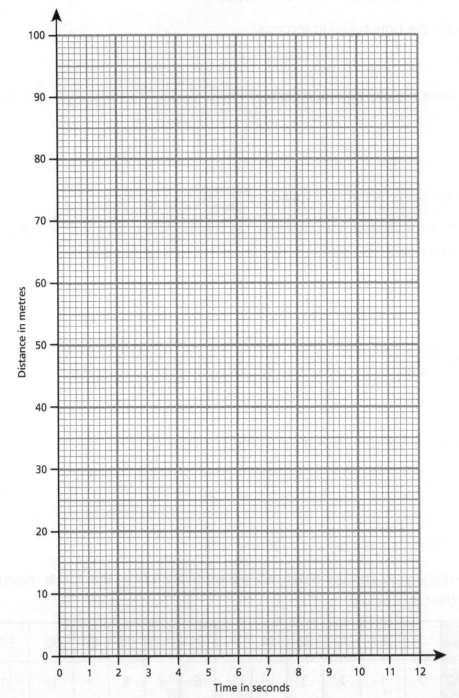

b) On **Graph 1**, draw a line through the points to show the motion of the athlete during the race. (Hint: the first few points, and the last few points, follow a curved line.) [1]

c) Calculate the athlete's average speed for the whole of the race. Show your working.

_____ m/s [2]

d) i) Determine the distance the athlete travels during the first second of the race.

.... m [1]

ii) Determine the distance the athlete travels between time = 1 s and time = 2 s.

.... m [1]

iii) Determine the distance the athlete travels between time = 6 s and time = 7 s.

.... m [1]

iv) Determine the distance the athlete travels between time = 11 s and time = 12 s.

.... m [1]

e) Use your answers to part **d)** to help you with the following questions.

i) What is happening to the athlete's speed during the first 3 seconds of the race? Tick **one** box.

increases ☐   decreases ☐   constant ☐   [1]

ii) What is happening to the athlete's speed during the last 3 seconds of the race? Tick **one** box.

increases ☐   decreases ☐   constant ☐   [1]

iii) What is happening to the athlete's speed between 6 and 9 seconds? Tick **one** box.

increases ☐   decreases ☐   constant ☐   [1]

f) Look carefully at your completed **Graph 1** above and decide which of the following statements are **true** and which are **false**.

i) If the slope of the line gets steeper, the athlete's speed is increasing.

.... [1]

ii) If the slope of the line becomes less steep, the athlete's speed is decreasing.

.... [1]

iii) If the slope of the line does not change, the athlete's speed is constant.

.... [1]

iv) If the slope of the line becomes less steep, the athlete's speed is increasing.

.... [1]

Total Marks .... / 28

Testing Understanding

1 **a)** A teacher is showing her pupils a spring with a mass attached to it, as seen in the diagram. There are two forces acting on the mass. There is an upward force on the mass due to the tension in the spring.

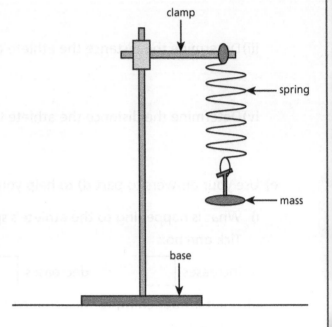

    i) Give the name of the downward force acting on the mass.

       _____ **[1]**

    ii) The mass attached to the spring is stationary. What does this tell us about the two forces acting on the mass?

       _____

       _____ **[1]**

b) The teacher now pulls the mass downwards towards the bench, stretching the spring further. She then releases the mass, which accelerates upwards very quickly.

Which statement correctly describes the two forces acting on the mass as it accelerates upwards? Tick **one** box.

The forces are in equilibrium. ☐

The upward force is greater than the downward force. ☐

The downward force is greater than the upward force. ☐ **[1]**

2 The diagram shows the horizontal forces on a delivery van moving along a flat, straight section of road.

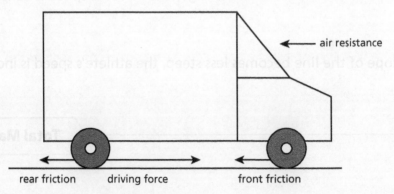

a) The forces on the van, while on this stretch of road, are shown in the table below:

Force	Driving force from engine	Air resistance	Front friction	Rear friction
Size of force (newtons)	2200	400	300	300

i) The forces on the van are unbalanced.

What is the size of the unbalanced force and is it in the forwards or backwards direction?

unbalanced force = _____ N [1]

direction = _____ [1]

ii) How would this unbalanced force affect the motion of the van? Tick **one** box. [1]

van speeds up ☐   van slows down ☐   van has a constant speed ☐

3 The amount of unbalanced force acting on an object is also known as the **resultant force**. Details of the size of the horizontal forces acting on some moving objects are shown in the table.

Moving object	Driving force (newtons)	Friction (newtons)	Air resistance (newtons)	Resultant force (newtons)	Direction of resultant
cyclist	0	5	15		
car	900	400	500		
bus	3000	600	700		

a) Complete columns 5 and 6 of the table by calculating the resultant and its direction for each of the moving objects listed. (Give the direction as either **forward**, **reverse**, or **none**.) [3]

b) Describe the effect the resultant has on the speed of each moving object.

i) Cyclist: _____ [1]

ii) Car: _____ [1]

iii) Bus: _____ [1]

4 The distance–time sketch graphs below represent the motion of four cars – A, B, C and D – on a straight, flat test track. At time = 0, the cars pass a marker on the track. Distance measurements are made from this marker.

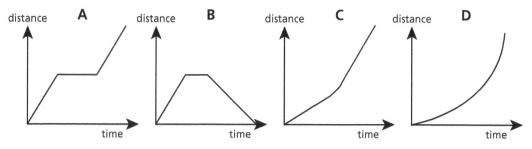

For the following questions, tick **one** box.

a) Which two cars were stationary at some stage in their motion?

A and B ☐   B and C ☐   C and D ☐   A and D ☐   [1]

b) Which car is accelerating for the whole of its motion?

A ☐   B ☐   C ☐   D ☐   [1]

c) Which car reverses at some stage in its motion?

A ☐   B ☐   C ☐   D ☐   [1]

d) Which car travels at a constant speed, accelerates briefly, then travels at a constant higher speed?

A ☐   B ☐   C ☐   D ☐   [1]

5 The distance–time graph represents the motion of two cyclists, Z and Y, along a straight, flat section of road. At time = 0, the cyclists pass a marker on the track. Distances are measured from this marker.

a) Determine the speed of cyclist Z in metres per second.

.... m/s   [1]

b) Determine the speed of cyclist Y in metres per second.

.... m/s   [1]

c) Determine the relative speed of the two cyclists.

.... m/s   [1]

d) The motion of the two cyclists continues unchanged and is monitored for a further 5 seconds. Calculate the distance between the cyclists at time = 10 seconds. Explain how you worked out your answer.

....

distance = ... m   [2]

6 The diagram below shows that the axis of rotation of the Earth is tilted.

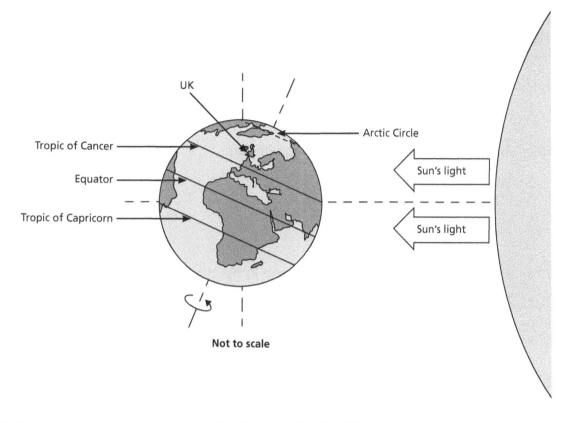

UK

Tropic of Cancer

Equator

Tropic of Capricorn

Arctic Circle

Sun's light

Sun's light

Not to scale

a) State what causes daytime and night-time in the UK.

_____ [1]

b) Explain what causes summer and winter in the UK.

_____

_____

_____ [2]

7 The image represents our galaxy, the Milky Way. The diameter of the Milky Way galaxy is about 106 000 light-years.

a) Define the term light-year.

_____ [1]

b) Explain why astronomers use the light-year as a distance unit instead of kilometres or miles.

_____ [2]

Total Marks _____ / 27

Working Scientifically

1 **a)** Explain how a stationary object can be in equilibrium.

.... [1]

b) Explain how a moving object can be in equilibrium.

....

....

.... [3]

c) Antonio is going to use the apparatus shown in the diagram below to demonstrate a stationary object in equilibrium under the action of four forces. The object is a metre rule. The apparatus is set up so that the metre rule is horizontal.

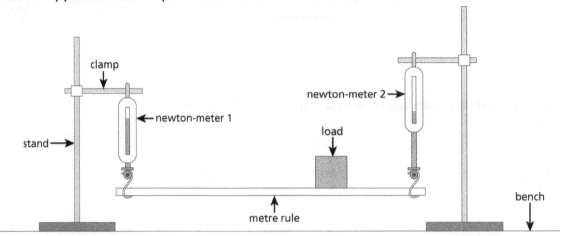

i) There are four vertical forces acting on the metre rule. Antonio records the size of the forces in the table below. Complete column 3 of the table to show the directions of the forces.

Force	Size of force (newtons)	Direction (up or down)
newton-meter 1	1.9	....
weight of metre rule	1.0	....
weight of the load	5.0	....
newton-meter 2	4.1	....

[4]

ii) Use the information in the table to explain how Antonio's measurements confirm what is meant by equilibrium. Show any working out that you do.

....

....

.... [3]

d) Another student, Shari, replaces the load on the metre rule with one that has a weight of 6.0 N. She adjusts the height of the newton-meters so that the metre rule is again horizontal.

i) Suggest how Shari could ensure that the metre rule is horizontal.

....

....

.... [3]

ii) Newton-meter 1 now reads 2.2 N. Use what you know about equilibrium to predict the reading on newton-meter 2. Show any working out that you do.

reading = ... N   [2]

Total Marks .... / 16

1 Read the passage below about the journeys of the Voyager 1 and the Voyager 2 spacecraft through the Solar System, and then answer the questions that follow.

Two NASA spacecraft, Voyager 1 and Voyager 2, have travelled beyond our Solar System on journeys taking over 40 years. The spacecrafts could only carry enough fuel to get them from Earth to Jupiter. Beyond Jupiter, with no more fuel, the gravitational force from the Sun slows down the spacecrafts. So how were Voyager 1 and Voyager 2 able to overcome the Sun's gravitational field and escape from the Solar System? Well, with careful planning and lots of brilliant mathematics, NASA was able

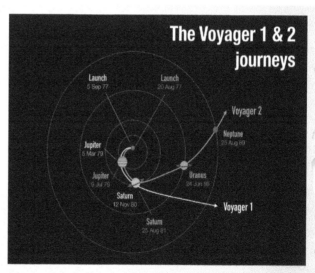

to plan the route of the spacecrafts so that the gravitational fields of certain planets dragged the spacecrafts along with them and speeded them up. This process is called a gravity assist manoeuvre or slingshot. The image shows the slingshots experienced by Voyager 1 and Voyager 2.

a) i) State the alternative name for a slingshot.

_____ [1]

ii) How many slingshots are shown in the above diagram for Voyager 1?

_____ [1]

iii) How many slingshots are shown in the above diagram for Voyager 2?

_____ [1]

b) i) What planet did the Voyager spacecraft reach just as they ran out of fuel?

_____ [1]

ii) Explain what would have happened to the Voyager spacecraft if the slingshots had not taken place.

_____

_____ [2]

c) The Voyager spacecraft have now left the Solar System and entered interstellar space within our galaxy. Give the name of our galaxy.

_____ [1]

Total Marks _____ / 7

	Vocabulary Builder	Maths Skills	Testing Understanding	Working Scientifically	Science in Use
Total Marks	/ 15	/ 28	/ 27	/ 16	/ 7

Vocabulary Builder

1 The image below shows waves being created by raindrops falling onto the water surface of a pond. Each wave created is made up of a series of crests and troughs. Complete the sentences using the words from the box.

superposition	scattering	longitudinal	energy	reflection	transverse

a) The waves in the image are _____ waves because they make the water surface move up and down. [1]

b) When the crest from one wave passes through a trough from another wave, they cancel each other. This effect is called _____. [1]

c) A leaf floating on the pond is moved up and down as the waves passes. This means that the waves transfer _____. [1]

d) When the water waves hit a hard smooth surface forming the edge of the pond, they experience _____. [1]

2 **a)** Answer the questions using the phrases from the box.

water wave	sound wave	light wave

i) Which wave travels at 300 000 000 m/s? _____ [1]

ii) Which wave can travel through a vacuum? _____ [1]

iii) Which wave is **not** a transverse wave? _____ [1]

b) Of the three waves given in the box in part **a)**, how many can transfer energy?

Tick **one** box.

1 ☐   2 ☐   3 ☐   [1]

3 The diagrams A, B, C and D, represent examples of what can happen when a ray of light is incident on a surface.

A

B

C

D

Draw a line from each diagram label to the process that is occurring.

Label	Process
A	Absorption
B	Specular reflection
C	Refraction
D	Diffuse reflection

[4]

4 Use the words from the box to complete the sentences about energy transfer.

temperature	chemical	conduction	radiation	light

a) Energy is transferred by _____ when a hot solid object is in contact with a cold solid object. [1]

b) The amount of thermal energy stored in an object depends on its _____ and its mass. [1]

c) Energy is transferred from the Sun to Earth across the vacuum of space by

.... and infrared ....   [2]

d) When a torch is switched on, ... energy is transferred to ....   [2]

Maths Skills

1 The table shows the energy available from 1 kilogram of four different fuels.

Fuel	Energy in megajoules per kilogram
wood	17
coal	24
petrol	45
hydrogen	142

a) Calculate how much energy is available from a car's full petrol tank if the mass of the petrol is 40 kg.

.... megajoules [1]

b) Which provides the most energy? Tick **one** box.

3 kg of wood ☐   2 kg of coal ☐   1 kg of petrol ☐   [1]

c) Some cars are being developed to be powered by hydrogen. One advantage of these cars is that they do not cause air pollution. Refer to the above table to suggest another advantage to hydrogen-powered cars compared with petrol-powered cars.

.... [1]

2 The energy transferred by an appliance in one second is known as its power. Power is measured in watts, where 1 watt = 1 joule per second. The table contains the power of three different appliances.

Appliance	Power (watts)
microwave	800
toaster	1000
kettle	2000

Which appliance transfers the most energy in the time stated? Tick **one** box.

microwave for 3 minutes ☐   toaster for 2 minutes ☐   kettle for 1 minute ☐ [1]

3 The cost of running a household appliance with a power of 1000 watts for 1 hour is 20 pence.

a) What would be the cost of having a 1000-watt electric heater switched on for 2 hours?

.... pence [1]

b) What would be the cost of having a 2000-watt electric heater switched on for 1 hour?

.... pence [1]

c) What would be the cost of having a 500-watt television switched on for 1 hour?

.... pence [1]

4 A 1 kilowatt heater in a room in a house has been switched on for 1 hour.

a) How many joules of energy does the heater transfer in 1 second?

.... J [1]

b) How many seconds are there in 1 hour?

.... s [1]

c) How many joules of energy are transferred in 1 hour?

.... J [1]

d) The electricity meter in the house records the energy being transferred. However, the meter is unable to display numbers as large as the answer to part **c)**. Instead, the meter measures energy transferred in different units. The meter counts '1' when a 1 kilowatt appliance has been switched on for 1 hour. The meter's unit is called a kilowatt-hour (kW hour).

i) How many kW hours of energy are transferred when a 2-kilowatt appliance has been switched on for 2 hours?

.... kW hour [1]

ii) Each kW hour of energy costs 20 pence.

What is the cost of having a 2 kW heater switched on for 2 hours?

.... pence [1]

5 An athlete in training is transferring 300 kJ of energy for every kilometre they run. The athlete runs for 2 hours at an average speed of 12 kilometres per hour.

a) What distance does the athlete run in kilometres?

.... km [1]

b) How much energy does the athlete transfer during their run? Give your answer in kilojoules.

.... kJ [1]

6 The diagram represents a machine for producing waves in a leisure pool. The machine has just been switched on.

direction wave travels

crest

wave machine
moves up and down

water surface

trough

An observer counts 6 crests passing a specific point in the pool in 1 minute.

Calculate the frequency of the wave in hertz.

.... Hz [1]

Total Marks .... / 15

Testing Understanding

1 a) Complete the sentences to describe the path taken by light that allows us to see objects outside during the day.

i) Light travels from the .... to Earth. [1]

ii) Light is .... by an object and enters our eyes. [1]

iii) Light is refracted by the eye's .... and lens. [1]

iv) An image of the object is formed on the .... [1]

v) Information about the image is sent to the brain by the .... [1]

b) Explain why it is more difficult to see objects outdoors at night.

.... [2]

2 The diagram shows a beam of white light incident on a glass prism.

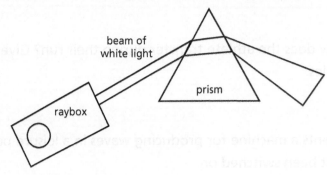

beam of white light

prism

raybox

a) What is the name of the process that causes the path of light to bend as it enters and leaves the prism?

.... [1]

b) Describe **two** ways in which the beam emerging from the prism is different to the beam entering the prism.

....

.... [2]

3 A T-shirt appears red when white light is shone on it.

Explain why the same T-shirt appears black when green light is shone on it.

....

.... [2]

4 Sylvie constructs a pinhole camera using an empty shoe box.

She uses a sharp pencil to make a hole at one end. At the other end she cuts out a square and covers it with greaseproof (wax) paper to form a screen.

Sylvie points the box at the candle. Viewing the screen at the end of the box, she sees the image of the flickering flame of the candle.

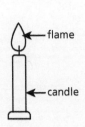

flame

candle

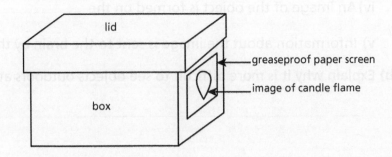

lid

greaseproof paper screen

image of candle flame

box

a) Draw two rays on the diagram below (one from the top and one from the bottom of the candle flame) that pass through the pinhole camera and hit the screen.

pinhole camera

flame

screen

[3]

b) Use your ray diagram to explain if the image of the flame is upright or upside down.

....

....

.... [2]

c) Sylvie then uses scissors to change the pinhole to a hole of diameter 2 cm. The image of the flame becomes much brighter but also very fuzzy and out of focus.

i) Explain why the image has become much brighter.

.... [1]

ii) Sylvie holds a lens in front of the hole. A bright, in-focus image of the flame and candle is formed on the screen.

Give the name of the type of lens she uses to focus the image.

.... [1]

iii) Give another example of a use of the type of lens named in part **ii)**.

.... [1]

5 A thermal mug can keep your coffee hot for several hours.

The thermal mug is a double walled steel container from which the air has been removed to create a vacuum.

vacuum

double wall
steel container

The thermal mug is designed to reduce the rate at which thermal energy is transferred from the hot liquid, inside the mug, to the surroundings.

a) Explain how the design of the mug affects energy transfer by conduction.

_____

_____ [2]

b) Explain how the design of the mug affects energy transfer by radiation.

_____ [1]

6 When events or processes occur, energy is transferred from one store to another, but no energy is actually lost. This is known as the Law of Conservation of Energy.

Consider the energy transfers taking place when a cyclist is cycling up a hill at a constant speed.

a) Tick the correct box (increasing, decreasing or constant) to show how the energy stores involved are affected.

Energy store	Increasing	Decreasing	Constant
kinetic			
gravitational potential			
chemical			
thermal			

[4]

b) Which one of the following correctly shows the transfer process between energy stores? Tick **one** box.

kinetic → gravitational potential + thermal ☐

chemical → gravitational potential + kinetic ☐

chemical → gravitational potential + thermal ☐

kinetic → gravitational potential + chemical ☐ [1]

Total Marks _____ / 28

1 Yulia is using a laser beam to demonstrate the reflection of light by a mirror.

a) What is the name given to this type of reflection?

.... [1]

b) The mirror is attached to a block of wood so that it can stand upright. Yulia stands the mirror on a piece of plain white paper on the bench. She draws a line along the front edge of the mirror as shown in **Diagram 1**.

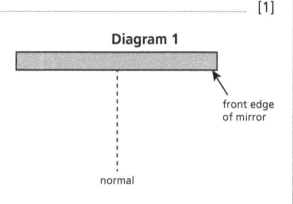

Diagram 1

front edge of mirror

normal

She then moves the mirror to one side and draws a dotted line at **90°** to the line just drawn. This line is called 'the normal'.

To draw the dotted line at 90°, Yulia uses the piece of equipment shown in **Diagram 2**.

Give the name of this piece of equipment.

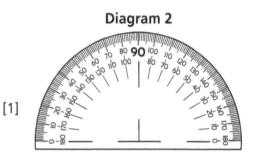

Diagram 2

.... [1]

c) Yulia sets up the apparatus as shown in **Diagram 3**. She is going to direct a red laser beam at the mirror. She plans to mark the incident and reflected rays so she can measure the angle of incidence, **I**, and the angle of reflection, **R**.

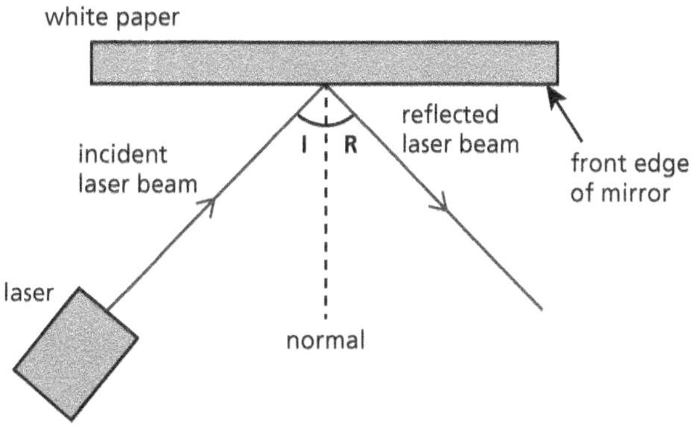

The beam of light from a laser can cause damage if it directly enters a person's eye. So before switching on the laser, the teacher explains to Yulia how to minimise the risk to herself and the other pupils in the room.

Diagram 3

white paper

incident laser beam

reflected laser beam

front edge of mirror

I   R

laser

normal

Suggest a safety procedure that Yulia should follow.

.... [1]

d) To record the path of the laser beam, Yulia decides to make ink dots along the length of the beam, before and after it hits the mirror. The laser beam is about 3 mm wide, so she tries to make the dots in the middle of the beam as shown in **Diagram 4**. She will then join the dots to record the path of the laser beam on the paper.

Diagram 4

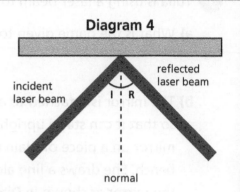

Yulia measures the angles: **I = 43°, R = 45°**

She was expecting to be able to demonstrate the law of reflection, which is:

angle of incidence I = angle of reflection R

i) Suggest why Yulia's values for the angle of incidence and the angle of reflection are not exactly the same.

....

.... **[1]**

ii) Suggest what Yulia should now do to demonstrate the law of reflection.

.... **[1]**

e) i) On **Diagram 5** below, use the equipment for angle measurement to draw a normal at the point where the incident ray hits the mirror.

Diagram 5

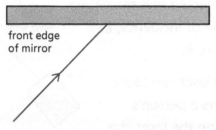

front edge of mirror

[1]

ii) Measure the angle of incidence and, on **Diagram 5**, draw the reflected ray so that angle R = angle I. **[2]**

Total Marks .... **/ 8**

1 Read the passage and answer the questions that follow.

Our sense of sight relies on the refraction of light. When light is incident on the eye, it is refracted by the cornea. The light then passes through the pupil into the lens, where it is again refracted. The lens can be squashed or stretched to change its shape and the amount of refraction it produces. The refraction of the light causes the formation of an image on the retina at the back of the eye. The cells on the retina are light sensitive and send information about the image along the optic nerve to the brain.

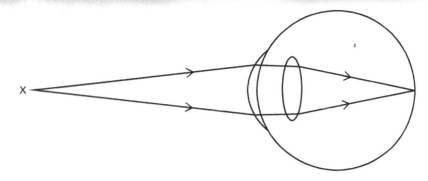

a) The simplified diagram above represents the refraction of light by a human eye.
   On the diagram above:

   i) label the names of the two parts of the eye where refraction of light occurs. [2]

   ii) label the name of the part of the eye that is photosensitive. [1]

b) i) The lens in the eye is made of a jelly-like substance. Suggest why this is the case.

   _____ [1]

   ii) What type of lens is in the eye?

   _____ [1]

c) It is important that the image formed on the retina is in focus. However, sometimes a defect of the eye means that the light is refracted too much or not enough, causing the image to be out of focus. State the usual way that such defects are corrected.

   _____ [1]

Total Marks _____ / 6

	Vocabulary Builder	Maths Skills	Testing Understanding	Working Scientifically	Science in Use
Total Marks	/ 18	/ 15	/ 28	/ 8	/ 6

The Periodic Table

Key

| relative atomic mass |
| **atomic symbol** |
| name |
| atomic (proton) number |

| 1 |
| **H** |
| hydrogen |
| 1 |

Group 1	Group 2												Group 3	Group 4	Group 5	Group 6	Group 7	Group 0
																		4 **He** helium 2
7 **Li** lithium 3	9 **Be** beryllium 4												11 **B** boron 5	12 **C** carbon 6	14 **N** nitrogen 7	16 **O** oxygen 8	19 **F** fluorine 9	20 **Ne** neon 10
23 **Na** sodium 11	24 **Mg** magnesium 12												27 **Al** aluminium 13	28 **Si** silicon 14	31 **P** phosphorus 15	32 **S** sulfur 16	35.5 **Cl** chlorine 17	40 **Ar** argon 18
39 **K** potassium 19	40 **Ca** calcium 20	45 **Sc** scandium 21	48 **Ti** titanium 22	51 **V** vanadium 23	52 **Cr** chromium 24	55 **Mn** manganese 25	56 **Fe** iron 26	59 **Co** cobalt 27	59 **Ni** nickel 28	63.5 **Cu** copper 29	65 **Zn** zinc 30		70 **Ga** gallium 31	73 **Ge** germanium 32	75 **As** arsenic 33	79 **Se** selenium 34	80 **Br** bromine 35	84 **Kr** krypton 36
85 **Rb** rubidium 37	88 **Sr** strontium 38	89 **Y** yttrium 39	91 **Zr** zirconium 40	93 **Nb** niobium 41	96 **Mo** molybdenum 42	[98] **Tc** technetium 43	101 **Ru** ruthenium 44	103 **Rh** rhodium 45	106 **Pd** palladium 46	108 **Ag** silver 47	112 **Cd** cadmium 48		115 **In** indium 49	119 **Sn** tin 50	122 **Sb** antimony 51	128 **Te** tellurium 52	127 **I** iodine 53	131 **Xe** xenon 54
133 **Cs** caesium 55	137 **Ba** barium 56	139 **La*** lanthanum 57	178 **Hf** hafnium 72	181 **Ta** tantalum 73	184 **W** tungsten 74	186 **Re** rhenium 75	190 **Os** osmium 76	192 **Ir** iridium 77	195 **Pt** platinum 78	197 **Au** gold 79	201 **Hg** mercury 80		204 **Tl** thallium 81	207 **Pb** lead 82	209 **Bi** bismuth 83	[209] **Po** polonium 84	[210] **At** astatine 85	[222] **Rn** radon 86
[223] **Fr** francium 87	[226] **Ra** radium 88	[227] **Ac*** actinium 89	[261] **Rf** rutherfordium 104	[262] **Db** dubnium 105	[266] **Sg** seaborgium 106	[264] **Bh** bohrium 107	[277] **Hs** hassium 108	[268] **Mt** meitnerium 109	[271] **Ds** darmstadtium 110	[272] **Rg** roentgenium 111								

Elements with atomic numbers 112–116 have been reported but not fully authenticated

*The Lanthanoids (atomic numbers 58–71) and the Actinoids (atomic numbers 90–103) have been omitted.

Cu and **Cl** have not been rounded to the nearest whole number.

Answers

Variation for Survival

Pages 4–18

Vocabulary Builder

1.

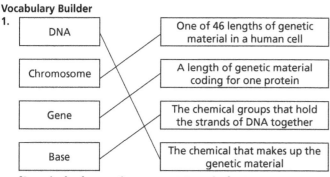

- DNA — The chemical that makes up the genetic material
- Chromosome — One of 46 lengths of genetic material in a human cell
- Gene — A length of genetic material coding for one protein
- Base — The chemical groups that hold the strands of DNA together

[3 marks for four or three correct; 2 marks for two correct; 1 mark for one correct]

2. a) true; false; false [3]
    b) false; true [2]
    c) true; false [2]
    d) true; true [2]

> Remember, natural selection is an ongoing process but usually happens very slowly.

3. environment **[1]**; mutations **[1]**; nucleus **[1]**; single **[1]**; two **[1]**; evolution **[1]**; survive **[1]**; characteristics **[1]**

Maths Skills

1. a) 7 [1]
    b) i)

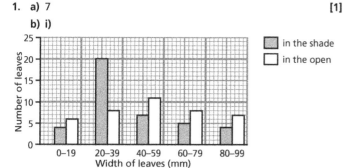

[2 marks for all five bars correctly plotted; 1 mark for only four correct bars]
    ii) 20–39 mm in the shade [1]
    iii) The leaves in the open are more evenly spread between groups; The leaves in the shade are not as wide. [2]
    c) Plant some of the plants from the shaded area in the open area **[1]**; Plant some of the plants from the open area in the shaded area **[1]**; See if the leaves grow narrower or wider **[1]**.

2. a) i) 5 times [1]
    ii) 24×0.3 **[1]**; $= 7.2\,kg$ **[1]**
    iii) A camel can lose more sweat without dying **[1]**; A camel can lose 80 kg of sweat **[1]**; A human can lose 8 kg of sweat **[1]**.
    b) i) more **[1]**; smaller than **[1]**
    ii) $\frac{6 \times 40}{10}$ **[1]**; $= 24\%$ **[1]**
    iii) Camel blood has a lower percentage of red blood cells **[1]**; In the desert, less water is available **[1]**; so a camel can drink less without the percentage of red blood cells becoming dangerously high **[1]**.

Testing Understanding

1. a) Woolly mammoth [1]
    b) Woolly mammoth → Giant moa → Tasmanian tiger → Polynesian tree snail [1]
    c)

- Snails can be born with different patterns of bands — Variation
- Different snails 'fight' each other for food — Competition
- Over many years the population of snails in a habitat changed — Evolution

[2 marks for two or three correct; 1 mark for one correct]

2. a) scar on cheek [1]
    b) cleft chin **[1]**; rounded nose **[1]**
    c) There are no intermediate types / it is either present or not. [1]
    d) i) They have different shaped chins and noses. [1]
    ii) One egg is fertilised by one sperm **[1]**; the embryo then splits into two **[1]**.

> Remember that if two eggs are released and each one is fertilised, then non-identical twins are produced.

3. a) They get one from each parent. [1]
    b) 47 [1]
    c) They have 3 chromosomes of 'pair' 21 / they have 47 chromosomes not 46. [1]
    d) They are male **[1]**; They have an X and a Y chromosome **[1]**.

4. a) circled: diagram B [1]
    b) She used X-rays to get data about the size and shape of DNA. ✓ [1]
    c) i) There is always the same percentage of G as C **[1]**; and the same percentage of T as A **[1]**.
    ii) They realised that the bases paired up G with C and A with T **[1]**; this pairing held the two chains of DNA together **[1]**.
    iii) $T = 15\%$; $G = 35\%$; $C = 35\%$ [3]

5. a) i) food [1]

> Remember that animals in different species do not successfully reproduce so they are not competing for mates.

    ii) **Accept one from:** The reds cannot digest the acorns very well; greys are more adapted to food source/habitat [1]
    b) A mutation might occur in the red squirrel that allows an individual to be able to digest acorns **[1]**; this makes the squirrel better adapted **[1]**; it is therefore more likely to survive **[1]**; and therefore, reproduce and pass on the mutation **[1]**.

Working Scientifically

1. a) To make it a fair test / to be able to compare the effect of light on the snails / control variable. [1]

> When using the term 'fair test', try to explain how it makes the experiment fair.

    b) The snails were in cages. [1]
    c) i) By seeing how faded the paint was **[1]**; the higher the number, the longer the snails stayed in the sunlight **[1]**.
    ii) The banded snails spent more time in the sunlight **[1]**; this is because they are better camouflaged **[1]**; so do not have to hide so much from predators **[1]**.

2. a) To act as a control [1]; so that they could see what effect the predator had on the lizards [1].
    b) The numbers of lizards on the islands with the introduced predator would decrease [1]; because more would be eaten by the predator [1]; The number of lizards on the islands without the predators would stay constant [1].
    c) i) The lizards with longer legs could run faster [1]; they would be less likely to be eaten [1]; they would be more likely to reproduce [1]; and pass on the genes for longer legs [1].
       ii) **Any two from:** The lizards were safer from predators in the trees; There was no advantage to having longer legs; Longer legs made it harder to climb trees [2]
    d) Natural selection usually takes a long time to happen. [1]

> The faster that organisms reproduce, the faster natural selection can occur – that is why it is often seen in bacteria.

Science in Use
1. a) A disease that is caused by a lack of a particular nutrient in the diet. [1]
    b) soft bones ✓ [1]
    c) $70 \div 6$ [1]; $= 11.7$ million [1]
    d) It stops an enzyme working [1]; therefore, provitamin D is not converted to cholesterol [1].
    e) To make sure that they are not harmful. [1]
    f) Because GE does not involve moving genes from one organism to another. [1]
2. a) $400\,000 \times 0.03 \div 100$ [1]; $= 120$ [1]
    b) Females have two copies of the X chromosome. ✓ [1]
    c) protein ✓ [1]
    d) Only boys have a Y chromosome. [1]
    e) If it is a boy then it will have a 50% chance of having muscular dystrophy [1]; they may decide to have a termination [1].

Our Health and the Effects of Drugs

Pages 19–27

Vocabulary Builder
1. nervous [1]; aspirin [1]; stimulants [1]; cocaine [1]; alcohol [1]; depressants [1]
2.
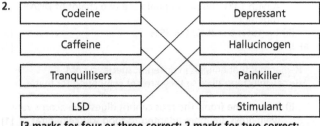

[3 marks for four or three correct; 2 marks for two correct; 1 mark for one correct]
3. a) true; false; true [3]
    b) false; false; true [3]
    c) true; false; true [3]

> Taking artificial steroids can reduce the levels of natural steroids (such as testosterone) in the body.

Maths Skills
1. a) $1.7 \times 10 = 17$ ml [1]
    b) $17 \div 440 \times 100$ [1]; $= 3.8636$ [1]; $= 3.9$ % [1]
    c) drinks $(1.8 \times 2) = 3.6$ units [1]; $= 216$ minutes [1]
    d) i) 4×0.02 [1] $= 0.08$ [1]
       ii) 14.5 times [1]

Testing Understanding
1. a) More dangerous drugs are put into higher categories [1]; There are stricter punishments for using/possessing more harmful drugs [1].

b) It causes the mind's perceptions to change. [1]
c) It means that they will respond more slowly in a dangerous situation [1]; so are more likely to have an accident [1].
d) It seems to show a correlation between cannabis and schizophrenia [1]; It does not mean that it causes it / there may be other factors linked to both cannabis and schizophrenia [1].

> Remember that a correlation between two factors does not mean that one causes the other.

2. a) crack cocaine [1]
    b) alcohol [1]
    c) methamphetamine; GHB [2]
    d) tobacco – passive smoking; alcohol – makes people more aggressive / may cause accidents if driving [2]
3. a) $1200 \div 50$ [1]; $= 24$ [1]
    b) It has the lowest dose that can kill [1]; There is the least difference between the dose that kills and the smallest dose that has an effect / smallest therapeutic ratio [1].

Working Scientifically
1. a) 13.6×3 [1]; $= 40.8$ or 41 patients [1]
    b) Non-smokers are very unlikely to get lung cancer. ✓; Compared to non-smokers, heavy smokers are more likely to die from other diseases as well as from lung cancer. ✓ [2]
    c) Doll studied many more patients. [1]

> Be prepared to answer questions about how information from studies like these would have been passed to other scientists or to the public in the different time periods.

d) There is a link between two factors [1]; it does not mean one factor causes the other [1].
e) Feature 1 – look at data from other hospitals; Feature 2 – look at the number of cigarettes people smoked; Feature 3 – check to see if people had cancer before they started smoking. [3]

Science in Use
1. a) People find it difficult to stop using it [1]; they have withdrawal symptoms without it [1].
    b) They cannot be sure how strong a batch of the drug is [1]; if it is very strong it could be harmful [1].
    c) Sativex contains THC [1]; which is hallucinogenic [1].
    d) Large doses could be harmful / people could become addicted to it. [1]
    e) It could be dangerous to test people [1]; do not want to introduce people to the drug [1].

Obtaining Useful Materials

Pages 28–38

Vocabulary Builder
1.
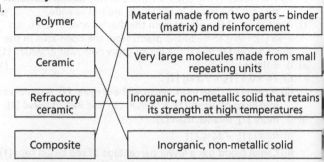

[3 marks for four or three correct; 2 marks for two correct; 1 mark for one correct]
2. absorb [1]; Breaking [1]; decomposition [1]; release [1]; Making [1]; combustion [1]

3. a) decomposition [1]
   b) displacement [1]
   c) smelting (**Accept:** displacement) [1]
   d) oxidation [1]
   e) displacement [1]
   f) decomposition [1]

   > Remember that all chemical reactions have a new substance that is produced.

4. a) ore ✓ [1]

   > Remember that minerals are compounds that contain metals. Ores are minerals that have such a high percentage of metal in them, that it is economical to extract the metal.

5. a) False [1]
   b) False [1]
   c) True [1]
   d) True [1]
   e) True [1]

Maths Skills

1. a) in a bar chart [1]
   b) Element [1]
   c) oxygen [1]
   d) 100 – 47 – 28 **[1]**; = 25% **[1]**
      Or (8 + 5 + 4 + 3 + 3 + 2) **[1]** = 25% **[1]**

   > Percentages mean 'per 100'. So, if you look at all the data, it adds up to 100% so this is all the elements found in the Earth's Crust.

2. a) volume [1]
   b) minute [1]
   c) 100 cm^3 [1]
   d) At the start there is no reaction and therefore no gas. **[1]** Then the reaction starts and there is a fast rate of reaction/speed at which the gas is produced, as the gas is made **[1]**. At about 2 minutes, the rate/speed of gas production slows down. **[1]**

3. percentage yield = (actual amount of product ÷ theoretical amount) × 100 **[1]** percentage yield = (0.9 ÷ 1.27) × 100 **[1]**; = 70.8661417 **[1]**; = 70.9% **[1]**

4. a) lead **[1]**; copper **[1]**
   b) About 500 years **[1]**
      (Accept an answer from 400–600 years)
   c) About 750 **[1]** BCE **[1]**
      (Accept an answer from 650 to 850 BCE)

Testing Understanding

1.

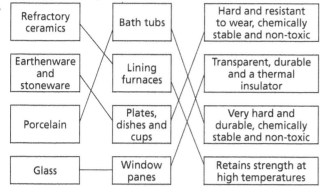

   [1 mark per each correct material linked to both use **and** property up to 4 marks]

   a) A list of metals **[1]** from most to least reactive **[1]**

   > The reactivity series can only be generated by observing the chemical reactions of the metals. It cannot be predicted.

   b) When a more reactive metal/element takes the place of/displaces a less reactive metal/element from its compound. **[1]**
   c) i) **Accept in either order**: calcium chloride **[1]** + zinc **[1]**
      ii) **Accept in either order**: magnesium sulfate **[1]** + copper **[1]**
      iii) magnesium nitrate + copper [1]
         (**Accept**: 'no reaction')
   d) Gold is the least reactive metal/at the bottom of this reactivity series. [1]
   e) To help make predictions on how the metal can be extracted from its ore **[1]**. Metals lower in the reactivity series than carbon can be extracted from their ore **[1]**.

3. a) monomers [1]
   b) starch [1]
      (**Accept**: cellulose)
   c) protein [1]
   d) Rubber is strong as it has a high number of chemical bonds in its structure **[1]**. Rubber is elastic as the long chains are tangled up **[1]** and straighten to long lengths when a force is applied **[1]**.

4. catalyst **[1]**; polyethene **[1]**; monomer **[1]**; crude oil **[1]**; polymer **[1]**; higher **[1]**; cross links **[1]**

5. a) smelting [1]
      (**Accept**: reduction/REDOX/displacement)
   b) **Accept in any order**: iron **[1]** + carbon dioxide **[1]**
   c) Carbon is more reactive than iron / higher than iron in the reactivity series **[1]**; so carbon can displace the oxygen from the iron oxide **[1]**.
   d)

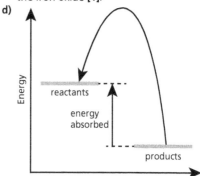

   [1 mark for correct *x*-axis label; 1 mark for correct *y*-axis label; 1 mark for reactants at a higher level than products; 1 mark for reaction pathway showing a 'hill']

6. a) Waste rock contains small amounts of toxic materials **[1]**. They can dissolve out of the rock over time/they can leach **[1]** and pollute the environment.
   b) Plant the mine area with trees and other plants **[1]**; and the hole caused by mining can be made into a lake **[1]**.
   c) Recycling avoids the need to mine and process new material **[1]** saving metal reserves **[1]**, so less energy is needed to recycle than to extract new metals from ores **[1]**. Less hazardous waste is produced from recycling **[1]**.

Working Scientifically

1. a) Bunsen burner [1]
   b) Gas tap [1]
      (**Accept:** gas cylinder/camping gas bottle)
   c) Top pan balance [1]
   d) 23.65 – 20.45 **[1]**; = 3.2 g **[1]**
2. a) Measuring cylinder [1]
      (**Accept:** burette or pipette)
   b) Temperature [1]
   c) Use a thermometer **[1]** to measure the temperature change **[1]**.

   > Valid means that the results allow you to answer the aim. In this case the observations need to allow Priya to decide if the reaction is exothermic or endothermic.

   d) If the temperature goes up, the reaction is exothermic **[1]**. If the temperature goes down, the reaction is endothermic **[1]**.
3. a) the metal [1]
   b) **Accept any three from**: same mass of metal; same surface area of metal; same type of acid; same concentration of acid; same temperature of acid [3]

c) Sodium is too reactive. **[1]**
(**Accept:** sodium would explode with the acid)
d) Use a textbook or suitable trusted website, e.g. Royal Society of Chemistry) **[1]**, to find out about the reaction of sodium with acid **[1]**.

Science in Use
1. **a)** Small stones **[1]**
   b) Construction **[1]**
   (**Accept:** building roads or houses)
   c) A material made from two parts **[1]**; a binder (matrix) and reinforcement **[1]**.
   d) Concrete fills the definition for a composite **[1]**. Concrete has a matrix in the form of cement **[1]** and a reinforcement in the form of stones **[1]**.
2. **a)** iron oxide **[1]**
   b) slag **[1]** iron **[1]**
   c) carbon + iron oxide → iron + carbon dioxide
   [1 mark for correct reactants; 1 mark for correct products; 1 mark for using →]
   d) Greenhouse effect/climate change **[1]**
   (**Accept:** acid rain)

Using our Earth Sustainably

Pages 39–48

Vocabulary Builder
1. **a)** False **[1]**
   b) False **[1]**
   c) True **[1]**
   d) True **[1]**
   e) True **[1]**
2.

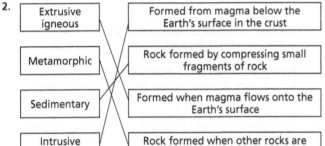

[3 marks if four or three correct; 2 marks if two correct; 1 mark if one correct]
3. atoms **[1]**; cycle **[1]**; footprint **[1]**; greenhouse **[1]**; carbon dioxide **[1]**; carbon dioxide **[1]**
4. **a)** **Accept in either order:** methane **[1]**; CFCs **[1]**
   b) **Accept in either order:** carbon dioxide **[1]**; methane **[1]**
   c) **Accept in either order:** sulfur oxide **[1]**; nitrogen oxide **[1]**
   d) nitrogen oxide **[1]**

Maths Skills
1. **a)** 100 − (78 + 1) = 21% **[1]**
   b) 21 : 78 **[1]**; 7 : 26 **[1]**

   Ratios should be expressed as the lowest values you can. So, after the ratio is written, look for common factors for all the numbers and use this to find the lowest possible numbers to express the ratio in.

   c) 120 × 0.78 **[1]** = 93.6 **[1]** = 94 g **[1]**
2. **a)** weeks **[1]**
   b) 26 000 − 260 **[1]**; = 25 740 **[1]** weeks
   c) 53 040 ÷ 52 **[1]**; = 1020 **[1]** years

   Remember, there are 52 weeks in a year.

   d) 2 weeks **[1]**; = 2 × 7 = 14 days **[1]**; × 24 = 336 hours **[1]**; × 60 = 20 160 seconds **[1]**; = 2.0×10^5 **[1]** seconds

3. **a)** **i)** 0.028% **[1]**   **ii)** 0.028% **[1]**   **iii)** 0.0315% **[1]**
   b) Just over 0.028% carbon dioxide in 1800 **[1]**; Just over 0.033% carbon dioxide in 1975 **[1]**; 0.033 − 0.028 = 0.005% **[1]**
   c) 0.028% is carbon dioxide so 100 − 0.028 = 99.972% is not carbon dioxide. **[1]**
   In a 100 g sample, 99.972 g would not be carbon dioxide **[1]**. This would round to 100% as the nearest percentage. **[1]**

   You have been asked to round to the nearest percentage, so look at the first decimal place and if the value is 5 or higher you must round up. This means that the calculation shows that 100% of the sample is not carbon dioxide. We know this is not true as we have measurements for the carbon dioxide and this shows how rounding can introduce error into data.

Testing Understanding
1.

Rock type / Feature	Igneous	Metamorphic	Sedimentary
Contains layers	✓	✓	
Most likely to have fossils			✓
Contains particles like sand and pebbles			✓
Contains crystals	✓	✓	

[1 mark for each correct tick up to a maximum of 6 marks]
2. **a)** A – Crust; B – Mantle; C – Outer core; D – Inner core
   [3 marks for four correct; 2 marks for three correct; 1 mark for two or one correct]
   b) **Accept in either order:** outer core **[1]**; mantle **[1]**
   c) **Accept in either order:** nickel **[1]**; iron **[1]**
   (**Accept:** Ni and Fe)
   d) **Accept in any order:** magnesium (oxide) **[1]**; silicon (oxide) **[1]**; iron (oxide) **[1]**
   e) Magma is molten material under the crust **[1]** and lava is the same molten material when it is above the crust **[1]**.
3. **a)** B **[1]**
   b) A **[1]**
   c) E **[1]**
   d) C **[1]**
   e) D **[1]**
4. **a)** Glass milk bottles can be collected and cleaned **[1]**. Fresh milk can be packaged in them **[1]** and the product re-sold **[1]**.
   b) Glass and plastic can be collected and separated into each type **[1]**. It can then be cleaned and melted down **[1]**, re-shaped and turned into a new product **[1]**.
   c) **Accept any four from:** down-cycling is when the product of the recycling process can't be recycled itself **[1]**; glass can be recycled an infinite number of times and its quality is not affected **[1]**; this means that products of recycled glass can be recycled themselves **[1]**; but, recycled PET often end up as products that they themselves cannot be recycled **[1]**; this reduces the sustainability of the PET compared to the glass **[1]**
5. **a)** False **[1]**
   b) True **[1]**
   c) True **[1]**
   d) False **[1]**
   e) True **[1]**
   f) True **[1]**

Working Scientifically
1. **a)** temperature **[1]**
   b) Wear eye protection **[1]**; Avoid getting any of the substance on her skin/wear gloves/wash hands after the experiment **[1]**
   c) Look at the crystals with a hand lens/microscope **[1]** and describe the size/sketch/take a photograph **[1]** with a scale/graticule in view **[1]**.
   d) Igneous (**Accept:** metamorphic) **[1]**

2. **a)** top pan balance [1]
   (**Accept:** balance)
   b) Accept any one from: nitric acid (**Accept:** HNO_3); sulfuric acid
   (**Accept:** H_2SO_4); sulfurous acid (**Accept:** H_2SO_3). [1]
   Accept suitable explanation, e.g. This acid is found in
   acid rain. [1]
   c) Accept any two from: same acid; same concentration of acid;
   same volume of acid; same mass of rock; same surface area
   of rock; same temperature of the reaction mixture [2]
   d) See bubbles **[1]**, hear fizzing **[1]** (**Accept:** effervescence **for
   2 marks**); mass decrease **[1]**.

Science in Use
1. **a) Accept in either order**: hydrogen **[1]**; carbon **[1]**
   b) Accept any one of: crude oil; coal; natural gas; shale gas [1]
   c) sulfur + oxygen → sulfur oxide/sulfur dioxide/sulfur trioxide
   **[1 mark for correct reactants; 1 mark for correct products;
   1 mark for using →]**
   d) acid rain [1]
2. **a) Accept in either order**: chalk **[1]**; limestone **[1]**
   b) Dead sea creatures with shells sink to the bottom of the sea/
   ocean **[1]**. Layers of other material fall onto **[1]** the dead
   sea creatures, which are then compacted **[1]** and cemented
   together **[1]**.
   c) carbonite **[1]**
   d) dolomite **[1]**

Motion on Earth and in Space

Pages 49–60

Vocabulary Builder
1. **a)** Earth [1]
   b) Sun [1]
   c) galaxy [1]
2. **a)** gravitational (force) (**Accept:** gravity) **[1]**; down **[1]**
   b) The forces are balanced **[1]**
   (**Accept:** forces cancel)
3. The car is accelerating so the forces acting are not balanced **[1]**
   and the car is not in equilibrium **[1]**.
   The cyclist has a constant speed, so the forces acting must be
   balanced **[1]**, so the cyclist is in equilibrium **[1]**.
   The bus is decelerating so the forces acting are not balanced **[1]**
   and the bus is not in equilibrium **[1]**.
4. light-year ✓ [1]
5. gravitational force [1]
   (**Accept:** gravity)
6. The Earth rotates on its axis from east to west. ✓ [1]

Maths Skills
1. **a)** weight = (80×10) = 800 N [1]
   b) i) $m = \frac{W}{g}$ ✓ [1]
   ii) mass = $\left(\frac{4}{10}\right)$ = 0.4 kg [1]
   iii) weight = (0.4×1.6) = 0.64 N [1]
2. **a) i)** distance = (1.5×1000) = 1500 m [1]
   ii) time = (2.5×60) = 150 s [1]
   iii) speed = $(1500 \div 150)$ = 10 m/s [1]

   There are 1000 metres in 1 kilometre, and 60 seconds in
   1 minute.

   b) i) time = $\frac{distance}{speed}$ ✓ [1]

   ii) time = $\left(\frac{8 \times 1000}{3.2}\right)$ = 2500 s [1]

   42 minutes **[1]** (**Accept more significant figures**)

3. **a)–b)**

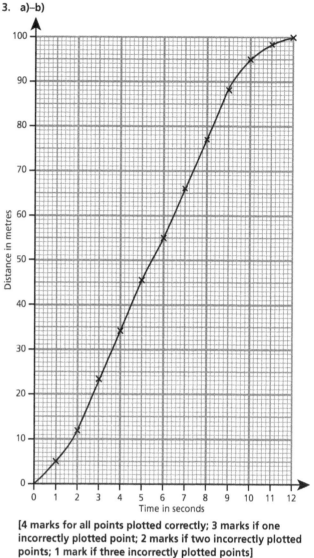

**[4 marks for all points plotted correctly; 3 marks if one
incorrectly plotted point; 2 marks if two incorrectly plotted
points; 1 mark if three incorrectly plotted points]**
**[1 mark for a single line through points, curved at beginning
and end]**
c) average speed = $\frac{100}{12}$ **[1]**
   8.3 m/s **[1]** (**Accept more significant figures**)

If you are asked to show your working when answering
a question using a formula, show the substitution of
numbers into that formula.

d) i) 5 m [1]
   ii) 7 m [1]
   iii) 11 m [1]
   iv) 2 m [1]
e) i) increases ✓ [1]
   ii) decreases ✓ [1]
   iii) constant ✓ [1]
f) i) True [1]
   ii) True [1]
   iii) True [1]
   iv) False [1]

Testing Understanding
1. **a) i)** weight [1]
      (**Accept:** gravitational force / gravity)
   ii) upward force = downward force [1]
      (**Accept:** tension = weight **or** The forces are in equilibrium.)
   b) The upward force is greater than the downward force. ✓ **[1]**
2. **a) i)** unbalanced force = (2200 – 400 – 300 – 300) = 1200 N **[1]**;
      forwards **[1]**
   ii) van speeds up ✓ [1]

3. a)

Moving object	Driving force (newtons)	Friction (newtons)	Air resistance (newtons)	Resultant force (newtons)	Direction of resultant
cyclist	0	5	15	20	reverse
car	900	400	500	0	none
bus	3000	600	700	1700	forward

[1 mark for each correct row up to a maximum of 3 marks]

b) i) Cyclist: speed decreases [1]
   (**Accept:** slows down / decelerates)
   ii) Car: constant/steady speed [1]
   iii) Bus: speed increases [1]
   (**Accept:** speeds up / accelerates)

4. a) A and B ✓ [1]
   b) D ✓ [1]
   c) B ✓ [1]
   d) C ✓ [1]

5.

> One way to determine speed from a straight line distance against time graph, is to use the graph axes to see how far the object travels between time = 0 and time = 1 s. This gives you distance travelled in 1 second, which is the speed. Check by getting the increase in distance for time = 1 to time = 2 s. However, if you have been taught how to calculate the gradient of a straight line then this will also give you the object's speed.

a) 8 m/s [1]
b) 5 m/s [1]
c) 3 m/s [1]
d) Cyclist Z travels 80 m in 10 s and Cyclist Y travels 50 m in 10 s [1]; distance = 30 metres [1]

6. a) The Earth's rotation [1]
   b) In the summer, the northern hemisphere tilts towards the Sun [1]. In the winter, the northern hemisphere tilts away from the Sun. [1]
7. a) The distance travelled by light in one year. [1]
   b) Astronomical distance/space distances are huge [1] so need a large unit [1].

Working Scientifically

1. a) The forces acting on the object are balanced. (**Accept:** no resultant force) [1]
   b) The object is moving in a straight line [1] at a constant speed [1] so the forces are balanced [1].
   c) i)

Force	Size of force (newtons)	Direction (up or down)
newton-meter 1	1.9	Up
weight of metre rule	1.0	Down
weight of the load	5.0	Down
newton-meter 2	4.1	Up

[1 mark for each correct entry in column 3 up to a max of 4 marks]

   ii) Total upward force = 1.9 + 4.1 = 6 N [1]; Total downward force = 1 + 5 = 6 N [1]; so total upward force = total downward force [1]
   d) i) Measure the height of the metre rule above the bench at different points [1]; adjust the height of clamps [1] to make the height of the metre rule above the bench constant [1].
   ii) newton-meter 2 reading = 6.0 + 1.0 − 2.2 [1]; 4.8 [1]

Science in Use

1. a) i) Gravity assist manoeuvre [1]
      ii) 2 [1]
      iii) 4 [1]
   b) i) Jupiter [1]
      ii) Spacecraft are slowed down [1] by the Sun's gravitational field [1]
   c) Milky Way [1]

Waves and Energy Transfer

Pages 61–71

Vocabulary Builder

1. a) transverse [1]
   b) superposition [1]
   c) energy [1]
   d) reflection [1]
2. a) i) light wave [1]
      ii) light wave [1]
      iii) sound wave [1]
   b) 3 ✓ [1]

3.

A	Absorption
B	Specular reflection
C	Refraction
D	Diffuse reflection

[1 mark for each correct line drawn to a maximum of 4 marks.]

4. a) conduction [1]
   b) temperature [1]
   c) light [1]; radiation [1]
   d) chemical [1]; light [1]

Maths Skills

1. a) energy = (40 × 45) = 1800 megajoules [1]
   b) 3 kg of wood ✓ [1]
   c) Hydrogen provides much more energy per kg. [1]
2. microwave for 3 minutes ✓ [1]
3. a) 40 pence [1]
   b) 40 pence [1]
   c) 10 pence [1]

> The total energy transferred by an appliance, and therefore the cost, depends on how quickly it transfers energy and how long it is switched on for.

4. a) 1000 joules [1]
   b) (60 × 60) = 3600 s [1]
   c) energy = (1000 × 3600) = 3 600 000 joules [1]
   d) i) energy = (2 × 2) = 4 kW hour [1]
      ii) cost = (4 × 20) = 80 pence [1]
5. a) distance = (2 × 12) = 24 km [1]
   b) energy = (24 × 300) = 7200 kJ [1]
6. frequency = (6 ÷ 60) = 0.1 Hz [1]

Testing Understanding

1. a) i) Sun [1]
      ii) reflected [1]
      iii) cornea [1]
      iv) retina [1]
      v) optic nerve [1]
   b) Lower light level/no sunlight [1]; so objects reflect less light into the eye [1].
2. a) refraction [1]
   b) **Any two from**: different direction of beam; beam is wider; beam contains different colours. [2]
3. The T-shirt only reflects red light [1]; The green light that was shone on the T-shirt is absorbed [1]. **Or:** As no light is reflected, the t-shirt will appear black. [2]
4. a)

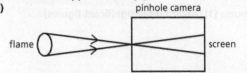

[1 mark for each correct ray, up to a maximum of 2 marks. 1 mark for having an arrow on at least one ray in the correct direction]

b) The image is upside down **[1]**; The ray from the top of the flame hits the screen at a lower position than the ray from the bottom of the flame **[1]**.

c) i) More light enters the box. **[1]**

   ii) convex **[1]**

   iii) Any one from: magnifying glass; spectacles; optical instruments; lens in an eye **[1]**

5. a) A vacuum/lack of air reduces energy transfer by conduction **[1]** because particles/solid needed for conduction. **[1]**

  b) Accept either: vacuum does not affect energy transfer by radiation **OR** energy transfer by radiation may be reduced by being reflected back into liquid by the steel container. **[1]**

6. a)

Energy store	Increasing	Decreasing	Constant
kinetic			✓
gravitational potential	✓		
chemical		✓	
thermal	✓		

[**1 mark for each correctly placed tick up to a maximum of 4 marks**]

  b) chemical → gravitational potential + thermal ✓ **[1]**

Working Scientifically

1. a) specular reflection **[1]**

  b) protractor **[1]**

  c) Any one of: do not look directly into the laser; do not point the laser at anyone; wear protective glasses; do not lift the laser above the bench **[1]**

  d) i) Any one of: difficulty in judging the middle of the laser beam; difficulty in judging the best fit line through the dots; random error **[1]**

    ii) Repeat the experiment. **[1]**

  e) i)

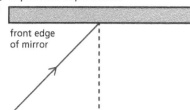

front edge of mirror

[**1 mark for normal drawn correctly**]

    ii)

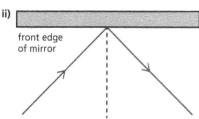

front edge of mirror

[**1 mark for drawing reflected ray so that angle R = angle I, and 1 mark for correctly putting an arrow on the reflected ray**]

When drawing a line to represent a ray of light, add an arrow head to the line to show the direction the light is travelling.

Science in Use

1. a) i)–ii)

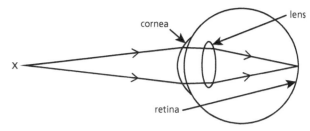

[**1 mark each for labelling cornea and lens up to 2 marks; 1 mark for labelling retina**]

  b) i) The lens has to be flexible/so it can be squashed or stretched **[1]**

    ii) Convex **[1]**

  c) Wearing spectacles/glasses **[1]**